Lecture Notes in Bioinformatics　　　12029

Subseries of Lecture Notes in Computer Science

More information about this series at http://www.springer.com/series/5381

Ion Măndoiu · T. M. Murali ·
Giri Narasimhan · Sanguthevar Rajasekaran ·
Pavel Skums · Alexander Zelikovsky (Eds.)

Computational Advances in Bio and Medical Sciences

9th International Conference, ICCABS 2019
Miami, FL, USA, November 15–17, 2019
Revised Selected Papers

Springer

Editors
Ion Măndoiu 🆔
University of Connecticut
Storrs, CT, USA

T. M. Murali
Virginia Tech
Blacksburg, VA, USA

Giri Narasimhan 🆔
Florida International University
Miami, FL, USA

Sanguthevar Rajasekaran 🆔
University of Connecticut
Storrs, CT, USA

Pavel Skums 🆔
Georgia State University
Atlanta, GA, USA

Alexander Zelikovsky 🆔
Georgia State University
Atlanta, GA, USA

ISSN 0302-9743   ISSN 1611-3349   (electronic)
Lecture Notes in Bioinformatics
ISBN 978-3-030-46164-5   ISBN 978-3-030-46165-2   (eBook)
https://doi.org/10.1007/978-3-030-46165-2

LNCS Sublibrary: SL8 – Bioinformatics

This Springer imprint is published by the registered company Springer Nature Switzerland AG
The registered company address is: Gewerbestrasse 11, 6330 Cham, Switzerland

Preface

The 9th edition of the International Conference on Computational Advances in Bio and medical Sciences (ICCABS 2019) was held during November 15–17, 2019, at Florida International University in Miami, Florida. ICCABS has the goal of bringing together researchers, scientists, and students from academia, laboratories, and industry to discuss recent advances on computational techniques and applications in the areas of biology, medicine, and drug discovery.

There were 30 extended abstracts submitted in response to the ICCABS 2019 call for papers. Following a rigorous review process in which each submission was reviewed by at least two Program Committee members, the Program Committee decided to accept 13 extended abstracts for oral presentation and one for poster presentation at the conference. The technical program of ICCABS 2019 included seven invited talks presented at the 9th Workshop on Computational Advances for Next Generation Sequencing (CANGS 2019) and five invited talks presented at the 2nd Workshop on Computational Advances for Single-Cell Omics Data Analysis (CASCODA 2019). Workshop speakers were invited to submit extended abstracts and, following the same review process used for the main conference, one additional extended abstract was selected for publication in this post-proceedings volume. All extended abstracts included in the volume have been revised to address reviewers' comments.

The technical program of ICCABS 2019 also featured keynote talks by four distinguished speakers: Prof. Yi Pan from Georgia State University gave a talk on "Biological Multiple Sequence Alignment: Scoring Functions, Algorithms, and Evaluations," Prof. Fangxiang Wu from University of Saskatchewan gave a talk on "Artificial Intelligence for Medical Image Analytics," Prof. Heng Huang from University of Pittsburgh gave a talk on "Large-Scale Machine Learning for Biomedical Data Science: AI Meets Health," and Prof. Shibu Yooseph from University of Central Florida gave a talk on "Identification of biomarkers and interactions from microbiome data." We would like to thank all keynote speakers and authors for presenting their work at the conference. We would also like to thank the Program Committee members and external reviewers for volunteering their time to review and discuss the submissions. Last but not least, we would like to extend special thanks to the Steering Committee members for their continued leadership, and to the finance, local arrangements, publicity, and publication chairs for their hard work in making ICCABS 2019 a successful event.

March 2020

Ion Măndoiu
T. M. Murali
Giri Narasimhan
Sanguthevar Rajasekaran
Pavel Skums
Alexander Zelikovsky

Organization

Steering Committee

Srinivas Aluru	Georgia Institute of Technology, USA
Reda A. Ammar	University of Connecticut, USA
Tao Jiang	University of California, Riverside, USA
Vipin Kumar	University of Minnesota, USA
Ming Li	University of Waterloo, Canada
Sanguthevar Rajasekaran (Chair)	University of Connecticut, USA
John Reif	Duke University, USA
Sartaj Sahni	University of Florida, USA

General Chairs

Giri Narasimhan	Florida International University, USA
Sanguthevar Rajasekaran	University of Connecticut, USA

Program Chair

T. M. Murali	Virginia Polytechnic Institute and State University, USA

Workshop Chairs

Ion Măndoiu	University of Connecticut, USA
Pavel Skums	Georgia State University, USA
Alex Zelikovsky	Georgia State University, USA

Finance Chair

Reda A. Ammar	University of Connecticut, USA

Local Arrangements Chairs

Giri Narasimhan	Florida International University, USA
Arpit Mehta	Florida International University, USA

Publicity Chairs

Orlando Echevarria	University of Connecticut, USA
Bob Weiner	University of Connecticut, USA

Publications Chair

Zigeng Wang   University of Connecticut, USA

Webmaster

Zigeng Wang   University of Connecticut, USA

Program Committee

Tatsuya Akutsu	Kyoto University, Japan
Max Alekseyev	George Washington University, USA
Jaime Davila	Mayo Clinic, USA
Jorge Duitama	Universidad de los Andes, Colombia
Scott Emrich	University of Tennessee, USA
Oliver Eulenstein	Iowa State University, USA
Liliana Florea	Johns Hopkins University, USA
Arnab Ganguly	University of Wisconsin, Whitewater, USA
Osamu Gotoh	National Institute of Advanced Industrial Science and Technology (AIST), Japan
Sumit Kumar Jha	University of Central Florida, USA
Yoo-Ah Kim	National Institute of Health, USA
Danny Krizanc	Wesleyan University, USA
M. Oguzhan Kulekci	Istanbul Technical University, Turkey
Manuel Lafond	Université de Sherbrooke, Canada
Yuk Yee Leung	University of Pennsylvania, USA
Ion Măndoiu	University of Connecticut, USA
Serghei Mangul	University of California, Los Angeles, USA
T. M. Murali	Virginia Tech, USA
Maria Poptsova	National Research University Higher School of Economics, Russia
Sanguthevar Rajasekaran	University of Connecticut, USA
Subrata Saha	IBM Corporation, USA
Pavel Skums	Georgia State University, USA
Yanni Sun	Michigan State University, USA
Sing-Hoi Sze	Texas A&M University, USA
Sharma V. Thankachan	University of Central Florida, USA
Mahmut Uludag	King Abdullah University, Saudi Arabia
Ugo Vaccaro	University of Salerno, Italy
Balaji Venkatachalam	Google, USA
Jianxin Wang	Central South University, China
Fang Xiang Wu	University of Saskatchewan, Canada
Shibu Yooseph	University of Central Florida, USA
Alex Zelikovsky	Georgia State University, USA
Shaojie Zhang	University of Central Florida, USA

Wei Zhang   University of Central Florida, USA
Cuncong Zhong   University of Kansas, USA

Additional Reviewers

Avdeyev, Pavel
Chahid, Abderrazak
Chakraborty, Dwaipayan
Icer, Pelin Burcak
Knyazev, Sergey
Kuzmin, Kiril
Markin, Alexey
Meade, Travis
Nekhai, Anton
Raj, Sunny
Srinivas, Srivathsan
Sun, Jiao
Tsyvina, Viachaslau

Contents

Detecting De Novo Plasmodesmata Targeting Signals and Identifying PD Targeting Proteins

Jiefu Li[1], Jung-Youn Lee[2,3], and Li Liao[1,3(✉)]

[1] Department of Computer and Information Sciences, University of Delaware, Newark 19716, USA
{lijiefu,liliao}@udel.edu
[2] Department of Plant and Soil Sciences, University of Delaware, Newark, DE 19716, USA
[3] Delaware Biotechnology Institute, University of Delaware, Newark, DE 19716, USA
lee@dbi.udel.edu

Abstract. Subcellular localization plays important roles in protein's functioning. In this paper, we developed a hidden Markov model to detect de novo signals in protein sequences that target at a particular cellular location: plasmodesmata. We also developed a support vector machine to classify plasmodesmata located proteins (PDLPs) in Arabidopsis, and devised a decision-tree approach to combine the SVM and HMM for better classification performance. The methods achieved high performance with ROC score 0.99 in cross-validation test on a set of 360 type I transmembrane proteins in Arabidopsis. The predicted PD targeting signals in one PDLP have been experimentally verified.

Keywords: Cellular localization · Support Vector Machines · Hidden Markov models

1 Introduction

It is well known that proteins after being synthesized have to be transported to their designated cellular location in order to fulfill the biological functions. However, much detail of the transporting mechanisms remain unknown, and subcellular localization prediction is an active research area in bioinformatics [1,10,14].

Plasmodesmata (PD) are membrane-lined intercellular communication channels through which essential nutrients and signaling molecules move between neighboring cells in the plant. This cell-to-cell exchange of molecules through PD is fundamental to the physiology, development and immunity of the plant and is a dynamically regulated cellular process. Several types of endogenous proteins, including type-I transmembrane proteins, as well as numerous PD-targeted

The work is funded by National Science Foundation NSF-MCB1820103.

I. Măndoiu et al. (Eds.): ICCABS 2019, LNBI 12029, pp. 1–12, 2020.
https://doi.org/10.1007/978-3-030-46165-2_1

proteins derived from plant viruses have been identified to associate with PD. However, no universal or consensus PD-targeting signal has ever been discerned nor molecular details are known as to how integral membrane proteins, including the best characterized PD-located proteins (PDLPs), are targeted to PD-specific membrane domains. As such, the current computational tools for cellular localization prediction do not even have PD categorized as a target location [1, 9].

So far, only eight PDLPs have been experimentally verified in Arabidopsis thaliana, and these proteins share a signature topology, as depicted in Fig. 1. Further experiments (unpublished) have narrowed PD targeting signal(s) down to a region, called extracellular juxtamembrane domain (JMe), which is between the DUF26 extracellular domain and the transmembrane domain (TMD). This region spanned about 20-amino acid residues, 9 AA downstream of the last conserved Cys residue of the DUF26 domain. Our experimental data (unpublished) pinpointed that the JMe region of PDLP5 contained a sufficient primary structure for its targeting to PD. Intriguingly, the data also implicated the presence of a second signal outside of the JMe region. However, the multiple sequence alignments of the eight Arabidopsis PDLP paralogs reveals no hints at the conserved amino acid residues or recognizable patterns shown in Fig. 2.

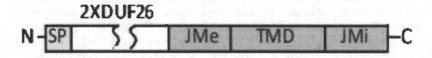

Fig. 1. Structure of PDLP

In this work, we set out to develop computational approaches to: i) detect subtle patterns that are associated with PD targeting, and ii) identify unknown PDLPs in Arabidopsis and other species. The second task can be considered as a classification problem, like other subcellular localization prediction problems. We adopted Support Vector Machine (SVM) [12] as a classifier and with dipeptide features to characterize proteins sequences. The performance of this straightforward approach is surprisingly good.

The first task of detecting PD targeting signal(s) turns out to be more challenging. Using the TMMOD tool [17] with customized training and feature selection, we detected some signals at the vicinity of the last conserved Cys of the DUF in addition to the initially defined JMe region. As for PDLP5, this prediction was consistent with the presence of a second signal. Further phylogenetic analysis of orthologues of PDLPs from other plant species and MEME motif finding [2] also identified similar but slightly stronger patterns in the same location for a subset of PDLPs (consisting of PDLP1 to PDLP4), as shown in Fig. 3.

These findings from the computational analyses together with the preliminary experimental data for PDLP5 prompted us to hypothesize that: there is the second PD-targeting signal outside of the JMe region might reside at the C-terminal end of the DUF domain. Unlike TMD, which has a clear-cut boundary at both ends from computational predictions, the C-terminal end of the DUF domain is not experimentally defined. This is why the previous experiments limited JMe as 20 AA adjacent to the N-terminal side of TMD, excluding 9 AA to avoid overstepping into the DUF domain. Based on the new hypothesis that this region likely contains a secondary signal for PD targeting, we newly define the JMe as the 30-AA region located between the rightmost conserved Cys at the very C-terminal end of the DUF domain and the N-terminal end of TMD.

Based on the aforementioned hypothesis, we then built a hidden Markov model (HMM) [3–7] to capture the functional structure of the JMe. The model has three states: state α for the left PD signal, state β for the right signal, and state γ for the non-functioning linkers. Using the trained hidden Markov model, we decoded the JMe region of the eight PDLPs, and the following-up experiments have verified the two PD signals and their relative positions in PDLP1, PDLP3, PDLP5 and PDLP8, as predicted by the model. Ongoing experiments are being conducted to verify predicted PD targeting signals for the remaining four PDLP proteins. Furthermore, the model was tested with predicting potential PDLPs in a dataset containing 360 type I transmembrane proteins, and showed remarkable performance as measured as ROC score in cross-validation.

Given the fact that PD targeting signals reside in JMe, we incidentally discovered a pitfall with the SVM classifier, when testing with randomized JMe to establish a baseline. SVM mistakenly classified these synthetic sequences – which are the same of the real PDLPs except for the JMe being randomized. To mitigate this issue, we propose a way to combine SVM and HMM to further improve the classification performance.

The paper is organized as follows. In Sect. 2, we describe in details the structure of the HMM, its training, and integration with SVM. In Sect. 3, we present the results on testing the hidden Markov model and SVM alone and the two methods in tandem. Conclusions are presented in the last section.

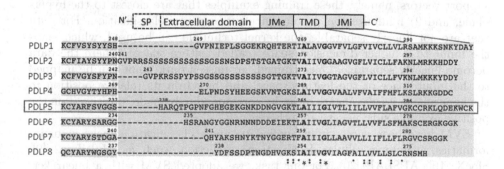

Fig. 2. Alignment of PDLP JMe, TMD, and JMc regions

```
PDLP1:  ▨▨▨▨▨▨▨▨▨▨▨▨▨ EPLSGGEKRQHTERT
PDLP2:  ▨▨▨▨▨▨▨▨▨▨▨▨▨ SSSSSSSSSSSSSSGSSNSDPSTSTGATGKT
PDLP3:  ▨▨▨▨▨▨▨▨▨▨▨▨▨ SSPYPSSGSSGSSSSSSSSGTTGKT
PDLP4:  ▨▨▨▨▨▨▨▨▨▨▨▨▨ SYHEEGSKVNTGKS
PDLP5:  ▨▨▨▨▨▨ VGGSHARQTPGPNFGHEGEKGN▨▨▨▨▨▨▨
PDLP6:  ▨▨▨▨▨▨ ARGGHSRANGYGGNRNN▨▨▨▨▨▨▨
PDLP7:  ▨▨▨▨▨▨ TDGAQHYAKSHNYKTNYGGERT
PDLP8:  ▨▨▨▨▨▨ GSGYYDFSSDPTNGDHVGKS
```

Fig. 3. MEME motifs discovered for PDLP JMe regions. The red region and green region are α state, β state in our HMM correspondingly. (Color figure online)

2   Methods

As mentioned in Introduction, we are tasked with detecting PD targing signals in the sequence of PDLPs and identifying potential novel PDLPs in Arabidopsis and other plant species. In this section, we describe in details the computational methods we developed for these two tasks. Our methods are based on two powerful machine learning methods – SVM and HMM. With the understanding the issues of using either SVM or HMM alone, we describe a way of combinating these two learning algorithms for better classification performance.

2.1   SVM with Dipepetide Features

SVM are a type of classifiers that take vectorized inputs and find the optimal separating hyperplane in the vector space with the positive training examples on one side of the hyperplane and negative training examples on the other side [13]. The optimization is to 1) maximize the margin between the hyperplane to the support vectors, namely, these training examples that are closest to the hyperplane, and 2) minimize the penalty incurred from misclassification. For data that are not linearly separable, the kernel technique can be used, which maps the input to a higher dimension space (called feature space), where the data become linearly separable. Once trained, an unseen data point can be mapped to the input space or the feature space, and based on its relative position to the hyperplane, its classification can be correspondingly made: positive if on the positive example side; negative if otherwise.

SVM have been successfully applied to many classification tasks in bioinformatics, including cellular localization prediction, such as MultiLoc2 [9], SlocX [19], APSLAP [20]. For our task, we adopted SVM with a linear kernel. Since the multiple sequence alignment (see Fig. 2) does not show clear high conservation patterns, we chose to use the alignment-free features to characterize the proteins for classification, specifically the dipeptide features [8]. In our case, the choice of dipeptide feature is a result of balancing the number of features and the number of training examples: dipeptide features capture more information than single amino acid composition and require less examples than tripeptide features and higher order k-mers, which have a dimension at 8000 and higher and require significantly more training examples.

The dipeptide features are the occurrence frequencies of all 400 possible amino acid pairs in a given protein sequence. Let $x_{dipeptide}$ denote the dipeptide feature, then $x_{dipeptide}(i, j)$ is the frequency of ith and jth amino acid as a neighboring pair to appear in the protein sequence. It is calculated as following:

$$x_{dipeptide}(i, j) = \frac{count([aa_i, aa_j])}{\sum_{i=1}^{20} \sum_{j=1}^{20} count([aa_i, aa_j])} \tag{1}$$

As a common practice, the dipeptide features are normalized as follows:

$$x_{dipeptide-norm} = \frac{x_{dipeptide} - \overline{x_{dipeptide}}}{std(x_{dipeptide})} \tag{2}$$

Although SVM with dipeptide features can be a powerful classifier, as shown in literature [9,19,20], it is worth noting that dipeptide occurrence frequency captures global features of the sequence as a whole and is hence not suitable for picking up subtle features from within short regions that are nevertheless important to the protein's functions. And this issue is exacerbated with insufficient amount of positive training data. In such cases, many dipeptide pairs may have zero counts in Eq. 1 – are these zeros real or will they become non zero should enough training examples be available? Consequently, SVM trained on these dipeptide features can be susceptible to overfitting and thus does not generalize well on unseen data.

In our case, we have only eight PDLPs and the JMe region that is known to contain PD-targeting signal is a very short region (around 30 amino acids long) as compared with the whole sequence (up to 700 amino acids long). As such, special attention should be paid to alleviate the aforementioned issues of training SVM with dipeptide features, for otherwise it may give rise to false positive predictions for the sequences that have similar dipeptide features but do not contain PD-targeting signals. Note that these issues are not unique only for SVM but for any classifiers that rely on features from full length sequences. For comparison, we also trained random forest classifier on dipeptide features, and the performances between the two classifiers are comparable, with SVM being slightly better.

2.2   3-States HMM on JMe

In contrast to the dipeptide approach of capturing more global information, the hidden Markov model we designed is focused on JMe region in order to detect PD-targeting signals that the web-lab mutagenesis experiments have suggested.

Based on the signature topology as shown in Fig. 1, the JMe region of a PDLP, or a potential PDLP such as type I transmembrane protein, can be easily extracted by two step procedure: 1) finding protein's transmembrane region via various sophisticated tools such as TMDOCK [18] and TMMOD [17], and 2) extracting 30 amino acids upstream of the transmembrane domain in step 1.

Our hidden Markov model has the 3 states, denoted as α, β, and γ. State α stands for PD-targeting signal A, and state β stands for PD-targeting signal B. State γ stands for everything else in the JMe region but the PD-targeting signals. The hidden states transition connections of these 3 states are shown in Fig. 4. The direct edge between state α and state β is to allow the case in which there are no linker residues between the two PD signals. Since we do not known for any given residue, which of the three states it is – in other words, the training data are unlabeled – we cannot train the HMM with counting as in a typical maximum likelihood approach, or with a multiple sequence alignment as in typical profile hidden Markov models for protein family classification. Instead, we adopted Baum-Welch algorithm, which is an expectation maximization approach and does not require the hidden states to be labeled in the training data [6].

After the model is trained, it is used for two tasks: 1) classifying PDLPs from a set of 360 type I proteins in Arabidopsis thaliana; 2) decoding the JMe region, i.e., marking out each residue as α, β or γ state.

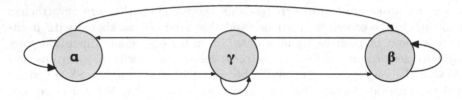

Fig. 4. 3-state HMM hidden states connections

It is worth to note that, because our 3-state HMM focuses on modeling with JMe, the model loses the global picture of PDLP sequences. For example, not every protein sequence with valid JMe region can be considered as PDLP. Furthermore, because HMM is not a discriminative model, the boundary between predicted positive and predicted negative is not readily given by the model and hence can be fuzzy. In the next subsection, we proposed a way to combine SVM and HMM, where we devised a mechanism to impose a threshold on HMM prediction score.

2.3   Combination of SVM with HMM

From the two previous subsections, we can see that SVM is good at capturing characteristic amino acid composition for full length sequences with dipeptide features, but less effective for short sequences like JMe region. This can potentially lead to false positive in predicting some protein sequences as PD targeting even though these proteins do not contain PD-JMe region. The issue becomes more apparent when testing the trained SVM to make de novo prediction of PDLPs in large dataset, in which more proteins may contain dipeptide features or even overall topology similar to that of the real PDLPs. On the other hand,

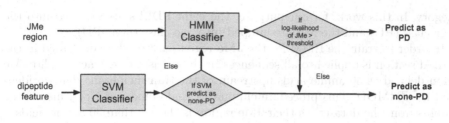

Fig. 5. Pipeline to combine SVM and HMM.

because the HMM focuses on the JMe region exclusively, it loses other features of PDLPs, e.g., it is found experimentally that a mutant without a proper JMi can not even be synthesized.

As such, the shortcomings of SVM and HMM both can give false positive predictions, but for different situations. The SVM tends to give false positive predictions for the sequences with none-PD JMe and PD like dipeptide feature, whereas HMM tends to give false positive predictions for the sequences with PD-JMe but none-PD's other domains because HMM only focuses on JMe region. Therefore, it is sensible to combine SVM and HMM in a complementary way that can overcome their shortcomings with their advantages. An option is to use decision tree on the predictions from SVM and HMM alone to choose a better one. However, due to the limited number of data samples, training decision tree by traditional method did not work. Fortunately, with the understanding of SVM and HMM in this particular task, we proposed a simple decision tree and the decision boundary for each node can be calculated properly, without use of a large training set.

The structure of combining SVM and HMM through decision tree to make final prediction is shown in Fig. 5. The node of SVM's decision boundary is naturally defined by the support vectors in the trained SVM. The decision boundary of HMM node is slightly tricky to define. Since we have the trained HMM model, synthetic JMe sequences can be generated. Given a large number of synthetic JMe sequences, their log-likelihoods follow a Gaussian distribution. By traditional statistic convention, 95% confidence interval is used, and we pick the lower bound of the interval as the decision boundary of HMM node.

3   Results

3.1   Datasets

Dataset A contains 360 type I membrane proteins in Arabidopsis thaliana, including the 8 PDLP sequences, all retrieved from Uniport [11]. From Uniport, labels of protein cellular localizations and their transmembrane domain are extracted, which account for 7 different types of membrane proteins, including endoplasmic reticulum membrane, Endosome membrane, Golgi membrane, Plasma membrane, Vacuole membrane, Vesicle membrane, and the PD membrane. Table 1 lists each cellular localizations and the number of proteins in that category. In this work, for our purpose, the eight PDLPs are grouped into the other class (PD) and the rest are lumped into one class (none-PD).

In order to train the HMM for the JMe region, a procedure described in the Method section is applied to all sequences in the dataset to extract a valid JMe region defined as 30 amino acids upstream of the transmembrane domain identified by TMMOD. This procedure eliminates Endosome and Golgi membrane proteins from the dataset, as their JMe region is shorter than 30 amino acids.

Table 1. Different types of proteins in dataset A

Protein localization	Endoplasmic reticulum	Plasma	PDLP	Vacuole	Vesicle
Number of proteins	17	322	8	7	6

To test the robustness of SVM and HMM in handling false positives, as described in the Method section, we add to the dataset eight synthetic sequences, which are the 8 PDLP sequences but with the origin JMe region being replaced with random residues. Since JMe region contains PD-targeting signal, it is highly confident that randomly replacing JMe region will lead to none-PD proteins. In other words, these eight synthetic sequences are negative data. Dataset A plus these eight synthetic sequences give rise to dataset B. Note that these eight synthetic sequences will be only used for testing.

As there are only 8 PDLP sequences, the leave-one-out (or equivalently 8-fold) cross validation scheme is adopted to ensure the maximum possible number of training examples to train the models. Specifically, each one of the 8 PDLP sequences is reserved as a positive test example once, and the remaining 7 PDLPs are used as positive training examples. The 352 none-PD sequences, as the negative examples, are randomly split into 8 subsets of equal size (44 sequences). One negative subset is picked to combine with the positive testing example to form the test set (45 sequences); and the remaining 7 negative subsets are merged together to form the negative training set. Note that, for the HMM, no negative training examples are needed. When dataset B is used, the whole process is the same, except that the synthetic none-PD sequences are repeatedly used as testing data for each fold.

3.2   Performance Metrics

To test the trained hidden Markov model $M(\theta)$, the test examples from the 8-fold cross-validation are combined and ranked by their prediction score $P(x|M(\theta))$, which is the likelihood for sequence x to be emitted from the model, calculated from the Forward algorithm. If a threshold is set for the prediction score, test examples with score above the threshold are classified as positive – they are true positive (TP) if their ground truth label is positive; they are false positive (FP) if otherwise. Similarly, test examples with score below the threshold are classified as negative – they are true negative (TN) if their ground truth label is negative;

they are false negative (FN) if otherwise. We use receiver operating characteristic (ROC) curve and ROC score, which is the area under the ROC curve, to evaluate the performance. The ROC curve plots the true positive rate against the false positive rate at the threshold sliding down the ranked list of test examples [15]. ROC curve starts (0,0) and goes to (1,1) in a monotonically manner. The perfect classifier has ROC score 1.0, and a random classifier has ROC score 0.5. When a natural choice of threshold is available, like the distance to the separating hyperplane in SVM, we also use the precision and recall associated with that threshold to evaluate the performance.

3.3   Evaluation: SVM Alone

In this experiment, we train and test a SVM with linear kernel on dipeptide features extracted from full length sequences. The ROC score from 8-fold cross-validation is 1.0 for dataset A but drops to 0.8837 for dataset B, because of misclassifying the synthetic none-PD as PD, which confirms the our concern that SVM alone can be susceptible to the overfitting issue. As comparison, the performance from a RF classifier is: ROC score = 0.9984 for dataset A and ROC score = 0.9177 for dataset B.

3.4   Evaluation: HMM Alone

In this experiment, we train the 3-state HMM as described in the method section. The trained HMM is then tested with 8-fold cross validation the ROC score is 0.93 for dataset A and is 0.9408 for dataset B. Unlike SVM, HMM's performance remains about the same for both datasets, confirming that HMM is more robust with false positives.

For the decoding task, the standard Viterbi algorithm [16] is used to scan the sequence against the model, trained with PDLP5 and ten of its orthologues, to annotate which residues belong to which of the three states, α, β or γ. So far, there is no ground truth available yet to directly evaluate the triple state annotation within JMe, except for PDLP5 and BAK1, the latter of which is experimentally confirmed as non PD targeting, see Fig. 6. It is very encouraging that the experiments for PDLP5 validated the existence of two PD targeting signals and their delineation is consistent with the annotation made by the model. More web-lab experiments are planned to validate model annotation of other PDLP paralogs.

3.5   Evaluation: Combination of SVM and HMM

From the experimental results in Subsects. 3.3 and 3.4, it is clear that SVM is susceptible to the pitfall of misclassifying the synthetic none-PD, whereas HMM is not affected. Also, by comparing the ROC score of SVM and HMM in dataset A, SVM has better performance and it naturally gives a clear decision boundary. In this experiment, we only focus on dataset B to show that by combining SVM

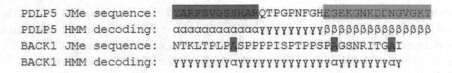

```
PDLP5 JMe sequence:   LARFSVGGSHAEQTPGPNFGHEGEKGNKDDNGVGKT
PDLP5 HMM decoding:   ααααααααααααααγγγγγγγγγγββββββββββββββββ
BACK1 JMe sequence:   NTKLTPLP SPPPPISPTPPSP GSNRITG I
BACK1 HMM decoding:   γγγγγγγγγαγγγγγγγγγγγγγγγαγγγγγγγγγαγ
```

Fig. 6. HMM decoding results for PDLP5 and BACK1. Red color and green color in the sequence represent state α, state β regions correspondingly. The region without color refers to state γ. (Color figure online)

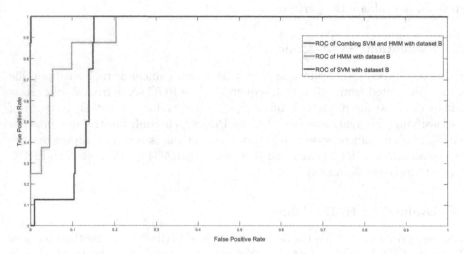

Fig. 7. ROC for combining SVM and HMM with dataset B (red), with ROC score = 0.9997, and comparing with using HMM alone (green), SVM alone (blue) (Color figure online)

with HMM via decision tree, performance can be improved, as compared to either SVM or HMM alone.

Figure 7 shows the ROC curve for the method of combining SVM with HMM (red) and the comparison with HMM (green) and SVM (blue) alone. The ROC score of combining SVM with HMM is 0.9997. Moreover, the precision/recall and ROC score comparison for all the three method for dataset B are shown in Table 2. For comparison, when SVM is replaced with RF, ROC score = 0.9169.

Table 2. Comparison for all method in dataset B

Method	ROC score	Precision	Recall
SVM with dipeptide feature	0.8837	0.1014	0.8750
HMM with JMe residues	0.9408	Not applicable	Not applicable
Combining SVM with HMM	0.9997	1.0000	0.8750

4   Conclusion and Future Work

In paper we presented computational approaches based on machine learning techniques to solve a very challenging biological problem: detecting plasmodesmata targeting signals and identifying novel PDLP sequences. The challenges arise from lack of clear sequence patterns, insufficient amount data, and unbalanced dataset. Without addressing these challenges, a straightforward application of standard machine learning techniques can lead to unreliable prediction, as demonstrated with using SVM on dipeptide features. In order to overcome these challenges, we closely incorporated domain specific knowledge into our hidden Markov model design, and devised a pipeline to leverage the predictive power of different models to reduce false positives. As a result, we are able to detect de novo PD targeting signals, verified by wet-lab experiments, and to classify PDLPs with remarkably high accuracy.

It is worth noting that, in this study, we adopted some common practices to avoid overfitting, such as the multi-fold cross-validation scheme and use of a simple linear kernel versus a more powerful kernel in SVM. While the performance from cross-validation as compared with training error does not indicate overfitting, given the small positive training examples in this study, it is difficult to know how well the trained classifiers will generalize to a large data set or data from different genomes, especially in detecting de novo PD proteins, which are actually being investigated in the web-lab experiments and no results to report yet. On the decoding task with our HMM, half of the predicted PD targeting signals have already been verified to be correct in the web-lab experiments, which are ongoing to verify the remaining predicted signals.

An online server will be deployed based the methods in the paper to assist biologists discovering new PDLP members. With new discovered PDLP members, an improved classifier can be built, which leads a positive feedback cycle of PDLP prediction and new PDLP members discovery. For the signal detection task, as the future work, the focus will be finding PD-targeting key residues in JMe region by extracting knowledge from the HMM to help with understanding the PD-targeting mechanism.

References

1. Almagro Armenteros, J.J., Sønderby, C.K., Sønderby, S.K., Nielsen, H., Winther, O.: DeepLoc: prediction of protein subcellular localization using deep learning. Bioinformatics **33**(21), 3387–3395 (2017)
2. Bailey, T.L., Elkan, C., et al.: Fitting a mixture model by expectation maximization to discover motifs in bipolymers (1994)
3. Baum, L.: An inequality and associated maximization technique in statistical estimation of probabilistic functions of a Markov process. Inequalities **3**, 1–8 (1972)
4. Baum, L.E., Eagon, J.A.: An inequality with applications to statistical estimation for probabilistic functions of Markov processes and to a model for ecology. Bull. Am. Math. Soc. **73**(3), 360–363 (1967)
5. Baum, L.E., Petrie, T.: Statistical inference for probabilistic functions of finite state Markov chains. Ann. Math. Stat. **37**(6), 1554–1563 (1966)

6. Baum, L.E., Petrie, T., Soules, G., Weiss, N.: A maximization technique occurring in the statistical analysis of probabilistic functions of Markov chains. Ann. Math. Stat. **41**(1), 164–171 (1970)
7. Baum, L.E., Sell, G.: Growth transformations for functions on manifolds. Pac. J. Math. **27**(2), 211–227 (1968)
8. Bhasin, M., Raghava, G.P.S.: ESLpred: SVM-based method for subcellular localization of eukaryotic proteins using dipeptide composition and psi-blast. Nucleic Acids Res. **32**(suppl_2), W414–W419 (2004)
9. Blum, T., Briesemeister, S., Kohlbacher, O.: MultiLoc2: integrating phylogeny and gene ontology terms improves subcellular protein localization prediction. BMC Bioinform. **10**(1), 274 (2009)
10. Chou, K.-C., Shen, H.-B.: Recent advances in developing web-servers for predicting protein attributes. Nat. Sci. **1**(02), 63 (2009)
11. The UniProt Consortium: UniProt: a worldwide hub of protein knowledge. Nucleic Acids Res. **47**(D1), D506–D515 (2018)
12. Cortes, C., Vapnik, V.: Support-vector networks. Mach. Learn. **20**(3), 273–297 (1995). https://doi.org/10.1007/BF00994018
13. Cristianini, N., Shawe-Taylor, J.: Support Vector Machines and Other Kernel-Based Learning Methods. Cambridge University Press, Cambridge (2004)
14. Dönnes, P., Höglund, A.: Predicting protein subcellular localization: past, present, and future. Genomics Proteomics Bioinform. **2**(4), 209–215 (2004)
15. Hanley, J.A., McNeil, B.J.: The meaning and use of the area under a receiver operating characteristic (ROC) curve. Radiology **143**(1), 29–36 (1982)
16. Heller, J., Jacobs, I.: Viterbi decoding for satellite and space communication. IEEE Trans. Commun. Technol. **19**(5), 835–848 (1971)
17. Kahsay, R.Y., Gao, G., Liao, L.: An improved hidden Markov model for transmembrane protein detection and topology prediction and its applications to complete genomes. Bioinformatics **21**(9), 1853–1858 (2005)
18. Lomize, A.L., Pogozheva, I.D.: TMDOCK: an energy-based method for modeling α-helical dimers in membranes. J. Mol. Biol. **429**(3), 390–398 (2017)
19. Ryngajllo, M., et al.: SLocX: predicting subcellular localization of Arabidopsis proteins leveraging gene expression data. Front. Plant Sci. **2**, 43 (2011)
20. Saravanan, V., Lakshmi, P.T.V.: APSLAP: an adaptive boosting technique for predicting subcellular localization of apoptosis protein. Acta Biotheor. **61**(4), 481–497 (2013). https://doi.org/10.1007/s10441-013-9197-1

The Agility of a Neuron: Phase Shift Between Sinusoidal Current Input and Firing Rate Curve

Chu-Yu Cheng$^{(\boxtimes)}$ and Chung-Chin Lu

Department of Electrical Engineering, National Tsing Hua University,
Hsinchu 30013, Taiwan
s9961829@m99.nthu.edu.tw, cclu@ee.nthu.edu.tw

Abstract. The response of a neuron when receiving a periodic input current signal is a periodic spike firing rate signal. The frequency of an input sinusoidal current and the surrounding environment such as background noises are two important factors that affect the firing rate output signal of a neuron model. This study focuses on the phase shift between input and output signals, and here we present a new concept: the agility of a neuron, to describe how fast a neuron can respond to a periodic input signal. By applying the score of agility, we are capable of characterizing the surrounding environment; once the frequency of periodic input signal is given, the actual angle of phase shift can then be determined, and therefore different neuron models can be normalized and compared to others.

Keywords: Integrate-and-fire model · Balanced background noise · Poisson process · Neuron model · Cellular modeling · Agility · Phase shift · Phase lag · Periodic signal · Sinusoidal current · Inter-spike interval · Firing rate · Membrane potential

1 Introduction

The integrate-and-fire (IAF) neuron model, a well-known spiking neuron model that has been studied for over a hundred years, becomes one of the most useful ways to analyze the behavior of neural networks, nervous systems, or even brain circuits [4,8,10,13,15]. Since the surrounding environment of a neuron *in vivo* is always noisy, each single neuron receives hundreds to thousands of excitatory and inhibitory postsynaptic potentials (EPSPs and IPSPs) within several seconds all the time, and the behavior of a neuron changes under such an environment [1,2,6,9,11,14]. A category of neuron models, called balanced leaky integrate-and-fire model (balanced LIF model) [3,4,7,16,17], includes this feature with additional parameters. A balanced LIF model contains two additional inputs to the membrane potential, that is, the excitatory background noise and the

© Springer Nature Switzerland AG 2020
I. Măndoiu et al. (Eds.): ICCABS 2019, LNBI 12029, pp. 13–25, 2020.
https://doi.org/10.1007/978-3-030-46165-2_2

inhibitory background noise. These two noise inputs are balanced in order to keep the membrane potential under the firing threshold when there is no current injection.

Because of the characteristics of the IAF neuron model, most of the important information such as temporal coding and rate coding can be retrieved through the study of a neuron's output spike train, in contrast, the waveform and peak height of a single action potential (spike) are relatively insipid [8]. The change of firing rate pattern over time can be described by an inter-spike-interval (ISI) firing rate function, which is often a smooth signal fluctuating in response to the fluctuation of the strength of the input signal. Although the IAF neuron model is not a linear time-invariant (LTI) system, with both of its input and output signals being periodic, similar approaches to investigate a LTI system can still be applied with some modifications [8,10]. An intuitive way to examine the dynamics of a LTI system is to measure the frequency response between periodic input and output signals. Frequency response is composed by amplitude response and phase response [10], here we only focus on the latter one.

The time constant is an important parameter within a differential equation describing the neuron membrane potential over time, governing how long it takes for the level of membrane potential to drop back to the resting state level after a fluctuation occurred. The value of time constant of a neuron model typically falls within the range between 10 and 100 ms [8]. However, the time constant is also a limitation of how fast a neuron can respond to an input stimulus and thus causes the shift or delay phenomenon in the phase response. This makes a large scale computational neuronal network containing more than tens of thousands of neurons or even with several hierarchical structures sluggish and cannot respond within a comparable time of a real nervous system *in vivo*. Previous studies [3,5] showed that a neuron can act in a more rapid way beyond its time constant limitation under background noise, indicating that the simple IAF neuron model lacks some important information and hence the balanced LIF model should be considered when investigating computational neuronal networks.

The remainder of this paper is organized as follows. First in Sect. 2, a balanced LIF model together with the sinusoidal current injection and background noise is constructed and the ISI firing rate function is then formulated. In Sect. 3, we derive the equation describing phase shift between input and output signals, and define the agility as one of the key factors that affect the phase shift. Relevant conclusions are drawn in Sect. 4.

2   Methods

Our way to present neuron firing rate statistics under periodic current injection can be learnt from Dayan and Abbott and Chance [7,8], and the balanced LIF model was originally referred to the Troyer and Miller's model [16]. The following subsection introduces the model we adopted.

2.1   The Balanced LIF Model

Troyer and Miller's Balanced LIF Model. The balanced LIF model proposed by Troyer and Miller is based on a typical *in vivo* cortical pyramidal cell [12]. The membrane potential in this model can be derived as follows

$$\tau_m \frac{dV}{dt} = -V + V_{rest} + g_{ex}(E_{ex} - V) + g_{in}(E_{in} - V) + V_{ext} \tag{1}$$

and

$$g_{ex} \rightarrow g_{ex} + n_{ex}d_{ex}$$
$$g_{in} \rightarrow g_{in} + n_{in}d_{in},$$

where g_{ex} and g_{in} are the excitatory and inhibitory synaptic conductances respectively, n_{ex} and n_{in} are the numbers of EPSPs and IPSPs received in each time step dt, and d_{ex} and d_{in} act as weighting factors that represent the resulting excitatory and inhibitory synaptic conductance changes. Between arrivals, both the excitatory and inhibitory conductances exponentially drop to zero governed by their own time constant τ_{ex} and τ_{in},

$$\tau_{ex} \frac{dg_{ex}}{dt} = -g_{ex}$$
$$\tau_{in} \frac{dg_{in}}{dt} = -g_{in}.$$

The values of parameters within (1) we applied here are listed in Table 1. The conductances g_{ex} and g_{in} are normalized by the input resistance of the neuron and expressed in dimensionless unit. The arrivals of the excitatory and inhibitory background noises are modeled as Poisson processes with rate λ_{ex} and λ_{in} respectively:

- Excitatory → 1000 Poisson inputs (N_{ex}), each with 7 Hz rate (λ_{ex})
- Inhibitory → 200 Poisson inputs (N_{in}), each with 15 Hz rate (λ_{in})

Table 1. Parameters of Troyer and Miller's balanced LIF neuron model

τ_m	V_{rest}	V_{th}	V_{reset}	τ_{ex}	τ_{in}	d_{ex}	d_{in}
50 ms	−74 mV	−54 mV	−60 mV	5 ms	10 ms	0.01	0.05

Sinusoidal Current Injection. Amplitude and frequency are the two parameters which can characterize a sinusoidal current injection. In this study, the input current is of the form: $I = I_0 + I_1 \cos \omega t$ where I_0 is the DC part and $I_1 \cos \omega t$ is the AC part of the current injection. The value of I_0 ranges from 1.0 nA to 2.0 nA, and I_1 from 0.1 nA to 0.5 nA. After the amplitude of the sinusoidal current injection is selected, different values of frequency ranging from 1 Hz to 1000 Hz are chosen in our experiment.

Background Noise Tuning. In order to balance the background noise beneath the firing threshold of the selected neuron model, there are several parameters we can adjust to meet the requirement: numbers of excitatory and inhibitory noises N_{ex} and N_{in}, Poisson rates λ_{ex} and λ_{in}, weights d_{ex} and d_{in}, time constants τ_{ex} and τ_{in}, and even the membrane resistance R_m, and specific membrane resistance r_m. Because even very tiny adjustment on $\lambda_{ex}, \lambda_{in}, d_{ex}, d_{in}, \tau_{ex}$, and τ_{in} will drastically change the behavior of our neuron model, we keep them untouched and only consider the remaining pair of parameters, i.e., the numbers of excitatory and inhibitory background noises (N_{ex}, N_{in}). The new parameter setting is shown in Table 2.

Table 2. Parameter setting of excitatory versus inhibitory background noises in our balanced LIF model.

Excitatory		Inhibitory
800	N	240
7	λ(Hz)	15
0.01	d	0.05
5	τ(ms)	10

2.2 ISI Firing Rate Function for Leaky IAF Model

In this subsection, we will begin with the derivation for ISI firing rate function.

The differential equation for the subthreshold membrane potential $V(t)$ between action potentials from a basic leaky IAF neuron model is of the form:

$$c_m \frac{d}{dt} V(t) = -\overline{g_L}(V(t) - E_L) + I_s(t)/A \tag{2}$$

with the reset membrane potential $V(0) = V_{reset}$ right after the action potential, i.e., at $t = 0$.

By the Laplace transform, we have

$$V(s) = \frac{V(0)}{(s + 1/\tau_m)} + \frac{E_L/\tau_m}{s(s + 1/\tau_m)} + \frac{I_s(s)/C_m}{(s + 1/\tau_m)}. \tag{3}$$

It can be seen that $V(t)$ is a low-pass filtered version of $I_s(t)$ and by the inverse Laplace transform, we have

$$V(t) = V_{reset} e^{-t/\tau_m} + E_L(1 - e^{-t/\tau_m}) + \frac{1}{C_m} \int_0^t d\tau I_s(\tau) e^{-(t-\tau)/\tau_m}. \tag{4}$$

When $I_s(t) = I_0 + I_1 \cos \omega t$, (4) becomes

$$V(t) = (E_L + I_0 R_m) - (E_L + I_0 R_m - V_{reset}) e^{-t/\tau_m}$$
$$+ I_1 R_m \left(\frac{\cos \omega t + \tau_m \omega \sin \omega t}{1 + \tau_m^2 \omega^2} - \frac{1}{1 + \tau_m^2 \omega^2} e^{-t/\tau_m} \right). \tag{5}$$

The interspike interval t_{isi} is the smallest solution of (5) by substituting $V(t)$ with V_{th}. When the frequency ω is not big, the instantaneous postsynaptic firing rate can be approximated as

$$r_{\text{isi}}(t) \approx \frac{1}{\tau_m} \left[\ln \left(\left[1 + \frac{\alpha + V_{\text{th}} - V_{\text{reset}}}{E_L + I_0 R_m + \beta - V_{\text{th}}} \right]_+ \right) \right]_+^{-1} \tag{6}$$

where

$$\alpha = \frac{I_1 R_m}{\sqrt{1 + \tau_m^2 \omega^2}} \left(\frac{1}{\sqrt{1 + \tau_m^2 \omega^2}} - \cos(\omega t - \theta(\omega)) \right)$$

$$\beta = \frac{I_1 R_m}{\sqrt{1 + \tau_m^2 \omega^2}} \cos(\omega t - \theta(\omega))$$

$$\theta(\omega) = \cos^{-1} \frac{1}{\sqrt{1 + \tau_m^2 \omega^2}}.$$

To add balanced background noise into our neuron model, we need to update the differential equation of the membrane potential $V(t)$ in (2). The balanced background noise leaky integrate-and-fire neuron model will be discussed in next subsection.

2.3 ISI Firing Rate Function for Balanced LIF Model

When we consider the background excitatory and inhibitory noises, the membrane potential $V(t)$ between action potentials becomes:

$$c_m \frac{d}{dt} V(t) = -\overline{g_L}(V(t) - E_L) + \frac{g_{\text{ex}}}{r_m}(E_{\text{ex}} - V(t)) + \frac{g_{\text{in}}}{r_m}(E_{\text{in}} - V(t)) + \frac{I_s(t)}{A} \tag{7}$$

again with $V(0) = V_{\text{reset}}$ right after the action potential, i.e., at $t = 0$.

The ISI firing rate now becomes

$$r_{\text{isi}}(t) \approx \frac{1}{\gamma - \tau_{\text{eff}} \ln(\delta + \epsilon \cos(\omega t - \theta_{\text{eff}}(\omega)))} \tag{8}$$

where

$$\tau_{\text{eff}} = \frac{\tau_m}{1 + g_{\text{ex}} + g_{\text{in}}}$$

$$\theta_{\text{eff}}(\omega) = \cos^{-1} \frac{1}{\sqrt{1 + \tau_{\text{eff}}^2 \omega^2}}$$

$$\gamma = \tau_{\text{eff}} \ln \left(E_L + g_{\text{ex}} E_{\text{ex}} + g_{\text{in}} E_{\text{in}} + I_0 R_m \right.$$

$$\left. + \frac{I_1 R_m}{1 + \tau_{\text{eff}}^2 \omega^2} - (1 + g_{\text{ex}} + g_{\text{in}}) V_{\text{reset}} \right)$$

$$\delta = E_L + g_{\text{ex}} E_{\text{ex}} + g_{\text{in}} E_{\text{in}} + I_0 R_m - (1 + g_{\text{ex}} + g_{\text{in}}) V_{\text{th}}$$

$$\epsilon = \frac{I_1 R_m}{\sqrt{1 + \tau_{\text{eff}}^2 \omega^2}}. \tag{9}$$

Equation (8) is our new ISI firing rate function for the balanced LIF neuron model. Note that in our balanced LIF model, the reversal potential of the excitatory background noise E_{ex} is set to be zero, so actually the term $g_{ex}E_{ex} = 0$ in (8) and (9) when applied within this study, and thus we only need to consider the γ and δ in (9) as

$$
\gamma = \tau_{eff} \ln \left(E_L + g_{in} E_{in} + I_0 R_m + \frac{I_1 R_m}{1 + \tau_{eff}^2 \omega^2} - (1 + g_{ex} + g_{in}) V_{reset} \right)
$$
$$
\delta = E_L + g_{in} E_{in} + I_0 R_m - (1 + g_{ex} + g_{in}) V_{th}. \tag{10}
$$

3   Results and Discussions

Figure 1 shows two situations that affect the phase. One is the adjustment of the frequency of the sinusoidal current injection f and the other is the different combinations on the number of background noise pair N_{ex} vs N_{in}. Figure 2 illustrates that the degree of phase shift can be expressed as a function of the (N_{ex}, N_{in}) pair under different frequencies of the sinusoidal current injection. As the total number of background noises increases, we can observe that the angle Φ of the phase shift diminishes accordingly. Also, higher frequencies of the sinusoidal current injection lead to larger angles of the phase shift.

The values of g_{ex} and g_{in} here in (8), (9), and (10) are substituted correspondingly by the average values $\overline{g_{ex}}$ and $\overline{g_{in}}$ recorded through the whole simulation course of 100 cycles with time step resolution $\Delta t = 0.01$ ms and 1000 ms per cycle, i.e. the average value in 10^7 points. (see Fig. 3(a))

3.1   The Angle of Phase Shift

Taking derivative of $r_{isi}(t)$ with respect to t yields

$$
\frac{d}{dt} r_{isi}(t) \approx \frac{d}{dt} \left(\frac{1}{\gamma - \tau_{eff} \ln \left(\delta + \epsilon \cos(\omega t - \theta_{eff}(\omega)) \right)} \right)
$$
$$
= \frac{\tau_{eff}\epsilon}{[\gamma - \tau_{eff} \ln \left(\delta + \epsilon \cos(\omega t - \theta_{eff}(\omega)) \right)]^2} \cdot \frac{-\omega \sin(\omega t - \theta_{eff}(\omega))}{\delta + \epsilon \cos(\omega t - \theta_{eff}(\omega))}. \tag{11}
$$

From (11) we can observe that, when $\sin(\omega t - \theta_{eff}(\omega)) = 0$, i.e. $\omega t = \theta_{eff}(\omega) + k\pi$ where $k \in \mathbb{Z}$ this derivative equals to zero. So the very first zero occurs at $t = \theta_{eff}(\omega)/\omega$. This value is only determined by the effective time constant τ_{eff} and the angular frequency ω of the sinusoidal current injection, and hence is only determined by $\overline{g_{ex}}$, $\overline{g_{in}}$ and ω. Note that the quantity of this value has nothing to do with the amplitude parameters of the current injection I_0 and I_1, which means that the degree of phase shift phenomenon is controlled by the characteristics of the background noise and the frequency of the sinusoidal current injection.

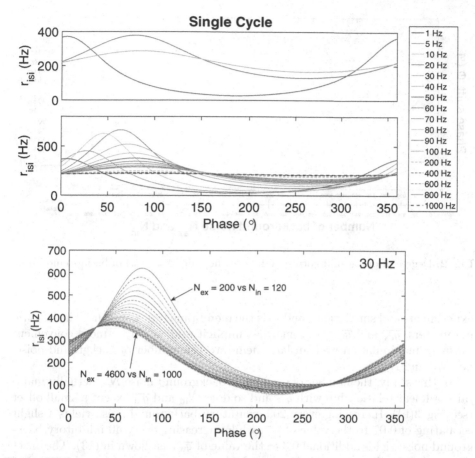

Fig. 1. ISI firing rate curve comparison. The amplitude of the sinusoidal current injection is fixed at $I_0 = 1.0\,\text{nA}$, $I_1 = 0.5\,\text{nA}$. (a) The r_{isi} curve undergoes phase shift when f increases from 1 Hz to 1000 Hz with the number of background noise pair fixed at $N_{\text{ex}} = 800$ vs $N_{\text{in}} = 240$. In the upper panel, we select three r_{isi} curves under $f = 1, 50, 100$ Hz representing lower frequency cases, while all different frequencies of the sinusoidal current injection ranges from 1 Hz to 1000 Hz we tested are plotted within the lower panel. (b) The r_{isi} curve undergoes phase shift with f fixed at 30 Hz while the number of background noise pair drops from $N_{\text{ex}} = 4600$ vs $N_{\text{in}} = 1000$ down to $N_{\text{ex}} = 200$ vs $N_{\text{in}} = 120$.

Consequently, once these four parameters: f, τ_m, $\overline{g_{\text{ex}}}$, and $\overline{g_{\text{in}}}$, are known, we can directly derive the degree of phase shift through the following equation

$$\Phi(°) = \arccos\left[1 + \left(\frac{2\pi f \tau_m}{1000(1 + \overline{g_{\text{ex}}} + \overline{g_{\text{in}}})}\right)^2\right]^{-1/2} \cdot \frac{360°}{2\pi}. \tag{12}$$

The first two parameters mentioned above are relatively intuitive for one to find out, since the frequency of the periodic current injection is always known in

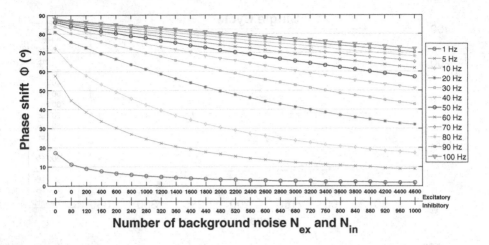

Fig. 2. Degree of phase shift corresponding to the pair $(N_{\text{ex}}, N_{\text{in}})$ of background noise.

experiments and simulations, and so is the membrane time constant. The last two parameters, $\overline{g_{\text{ex}}}$ and $\overline{g_{\text{in}}}$, are somewhat implicit and would be more convenient for us to manipulate if we translate them into the number of background noise, i.e. N_{ex} and N_{in}.

In this study, the number of excitatory background noise N_{ex} actually maintains a linear relationship with $\overline{g_{\text{ex}}}$ and so does N_{in} and $\overline{g_{\text{in}}}$ except a small offset (see Fig. 3(b)). Increasing every 200 excitatory background noise yields a slight escalating of 0.07 to the value of $\overline{g_{\text{ex}}}$, while increasing every 40 inhibitory background noise yields additional 0.3 to the value of $\overline{g_{\text{in}}}$, as shown in (13). The offset observed at inhibitory case is because the lowest pre-set pair of our background noise is starting from $N_{\text{ex}} = 200, N_{\text{in}} = 120$, that is, if we further reduce the number of background noise from the lowest pre-set pair to make $N_{\text{ex}} = 0$, then $N_{\text{in}} = 80$ instead of zero in our study.

$$\overline{g_{\text{ex}}} = \frac{0.07}{200} N_{\text{ex}}$$
$$\overline{g_{\text{in}}} = \frac{0.3}{40} N_{\text{in}}. \tag{13}$$

Applying the linear relationship described in (13) to the denominator within the parentheses in (12), the denominator becomes

$$1000(1 + \overline{g_{\text{ex}}} + \overline{g_{\text{in}}}) = 1000 + 0.35N_{\text{ex}} + 7.5N_{\text{in}}. \tag{14}$$

The coefficients of N_{ex} and N_{in} are actually product of the rest of the Poisson parameters λ, d, and τ, so (12) can be written in a more general form

$$\Phi(^\circ) = \arccos\left[1 + \left(\frac{2\pi f \tau_m}{1000 + p_{\text{ex}} + p_{\text{in}}}\right)^2\right]^{-1/2} \cdot \frac{360^\circ}{2\pi} \tag{15}$$

where

$$p_{ex} = N_{ex}\lambda_{ex}d_{ex}\tau_{ex}$$
$$p_{in} = N_{in}\lambda_{in}d_{in}\tau_{in}.$$

The range of Φ is $0° \leq \Phi \leq 90°$. When the two products of Poisson parameters p_{ex} and p_{in} are very large and the frequency of current injection f is the smallest value, i.e. $f = 1$, the value of Φ approaches to $0°$ with small τ_m. If background noise is small enough to be neglected or just simply equals to zero, then with very large frequency f and membrane time constant τ_m, the value of Φ approximately equals to $90°$.

3.2   The Agility Score of a Neuron

Here, we introduce a new concept in order to better describe the neuron's behavior responding to the surrounding environment factors: the agility of a neuron. Inspired by (12), the agility of a neuron is a function $AG(\tau_m, \overline{g_{ex}}, \overline{g_{in}})$ that counters the frequency of the periodic current injection. The AG function is defined as

$$AG(\text{Hz}) \triangleq \frac{1000(1 + \overline{g_{ex}} + \overline{g_{in}})}{2\pi\tau_m}, \tag{16}$$

and in this study, the agility of a neuron $AG(\tau_m, p_{ex}, p_{in})$ is of the form

$$AG(\text{Hz}) \triangleq \frac{1000 + p_{ex} + p_{in}}{2\pi\tau_m}. \tag{17}$$

From (16) and (17) we can see that the unit of agility score AG is Hz, which is identical to the frequency of the periodic current injection f. Neurons with higher AG-score can respond to a high frequency input more rapidly, and hence is more agile comparing to lower AG-score ones.

Now look back to (12), it can be rewritten as a function of f and AG

$$\Phi(°) = \arccos\left[1 + \left(\frac{f}{AG}\right)^2\right]^{-1/2} \cdot \frac{360°}{2\pi}$$

$$= \arccos\left(\frac{AG}{\sqrt{AG^2 + f^2}}\right) \cdot \frac{360°}{2\pi}. \tag{18}$$

The relationship among Φ, AG, and f is shown in Fig. 4. With this plot, one can easily set up any criteria and then derive desired conditions. For example, when the angle of phase shift constrained within only $5°$ under current injection $f = 1$ Hz is preferable in one's consideration, then the AG-score should be at least greater than 15 to guarantee such criteria; yet the AG-score must be at least greater than 1150 to satisfy the same criteria under current injection $f = 100\,\text{Hz}$,

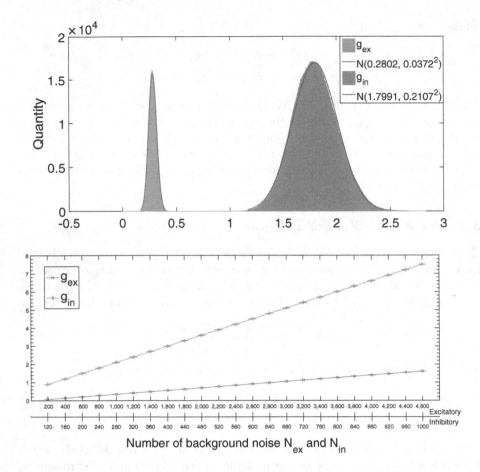

Fig. 3. (a) The distribution of g_{ex} and g_{in} under the background noise pair $N_{ex} = 800$ vs $N_{in} = 240$ can be fitted by normal distributions. There are 10^7 points of g_{ex} and g_{in} each, taken from one single arbitrary simulation course consisted of 100 cycles with time step resolution $\Delta t = 0.01$ ms and 1000 ms per cycle. (b) The average values of g_{ex} and g_{in} corresponding to pairs of background noise.

and when the frequency of sinusoidal current injection raised up to 1000 Hz, the AG-score now need to be greater than 11440 in order to keep the phase shift within $5°$.

The tolerance of the angle of phase shift can also be set as the proportion of one cycle alternatively, regardless of real time unit, for some simulation purposes. For example, if the level of phase shift is to be controlled within 0.1% per cycle, that is $\Phi \leq 3.6°$, then under $f = 1$ Hz of current injection, the AG-score should be greater than 160; while under $f = 50$ Hz, the demanding AG-score increased up to at least 7960, and AG-score should be over 15920 to satisfy the same criteria under $f = 100$ Hz of current injection. On the other hand, if the frequency of the sinusoidal current injection is fixed at $f = 1000$ Hz, then the AG-score

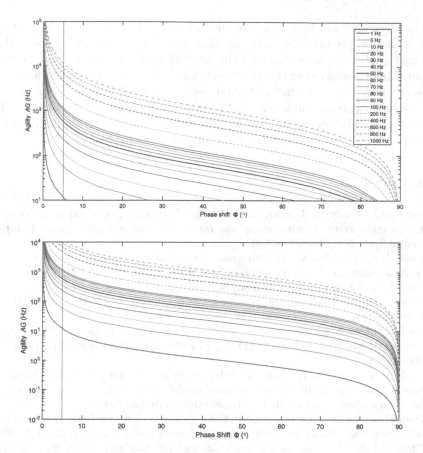

Fig. 4. (a) The AG scores corresponding to phase shift degree under different frequencies of current injection. (b) A further zoom-in at smaller AG values.

needs to be up to 57300 or above to keep the angle of phase shift smaller than $1°$, while it only takes $AG \geq 2870$ to keep the same criteria at the situation that the sinusoidal current injection frequency drops down to $f = 50\,\text{Hz}$.

4   Conclusion

The response of a neuron model to an external periodic stimulus can be affected by two summarized factors, one is the active role f, and the other the passive role AG. In this study, we present the agility function AG for a balanced LIF model with Poisson distributed background noise in the form of (17). This concept can be further generalized under the form of (16) and applied onto the same kind of neuron models like adaptive neuron models, or even onto conductance-based neuron models such as Hodgkin-Huxley models, Connor-Stevens models, etc. Previous studies also reported similar results [3,5] between background noise and the phase shift but were lack of quantitative descriptions. The novelty of this study is that we present an explicit function as a tool for us to directly calculate the exact degree of phase shift and the method of derivation of the equations presented here is relatively easy and straightforward for further applications. The AG score allows us to normalize various neuron models from different studies and makes them comparable with each other under the same conditions. This result also provides an explanation for how large-scale computational neuronal networks are able to overcome the input-output delay problem.

References

1. Amit, D.J., Tsodyks, M.: Quantitative study of attractor neural network retrieving at low spike rates: I Substrate-spikes, rates and neuronal gain. Netw. Comput. Neural Syst. **2**(3), 259–273 (1991)
2. Amit, D.J., Tsodyks, M.: Effective neurons and attractor neural networks in cortical environment. Netw.: Computat. Neural Syst. **3**(2), 121–137 (1992)
3. Brunel, N., Chance, F.S., Fourcaud, N., Abbott, L.: Effects of synaptic noise and filtering on the frequency response of spiking neurons. Phys. Rev. Lett. **86**(10), 2186 (2001)
4. Burkitt, A.N.: A review of the integrate-and-fire neuron model: I. Homogeneous synaptic input. Biol. Cybern. **95**(1), 1–19 (2006). https://doi.org/10.1007/s00422-006-0068-6
5. Burkitt, A.N., Meffin, H., Grayden, D.B.: Study of neuronal gain in a conductance-based leaky integrate-and-fire neuron model with balanced excitatory and inhibitory synaptic input. Biol. Cybern. **89**(2), 119–125 (2003)
6. Carandini, M., Mechler, F., Leonard, C.S., Movshon, J.A.: Spike train encoding by regular-spiking cells of the visual cortex. J. Neurophysiol. **76**(5), 3425–3441 (1996)
7. Chance, F.S.: Modeling cortical dynamics and the response of neurons in the primary visual cortex: Frances S. Chance. Ph.D. thesis, Brandeis University (2000)
8. Dayan, P., Abbott, L.F., Abbott, L.: Theoretical Neuroscience: Computational and Mathematical Modeling of Neural Systems. MIT Press, Cambridge (2001)
9. Gerstner, W.: Population dynamics of spiking neurons: fast transients, asynchronous states, and locking. Neural Comput. **12**(1), 43–89 (2000)
10. Izhikevich, E.M.: Dynamical Systems in Neuroscience. MIT Press, Cambridge (2007)
11. Knight, B.W.: Dynamics of encoding in a population of neurons. J. Gen. Physiol. **59**(6), 734–766 (1972)
12. McCormick, D.A., Connors, B.W., Lighthall, J.W., Prince, D.A.: Comparative electrophysiology of pyramidal and sparsely spiny stellate neurons of the neocortex. J. Neurophysiol. **54**(4), 782–806 (1985)
13. Rinzel, J.: Models in neurobiology. In: Enns, R.H., Jones, B.L., Miura, R.M., Rangnekar, S.S. (eds.) Nonlinear Phenomena in Physics and Biology. NATO Advanced Study Institutes Series (Series B: Physics), vol. 75, pp. 345–367. Springer, Boston (1981). https://doi.org/10.1007/978-1-4684-4106-2_9
14. Spekreijse, H., Oosting, H.: Linearizing: a method for analysing and synthesizing nonlinear systems. Biol. Cybern. **7**(1), 22–31 (1970)

15. Tranquillo, J.V.: Quantitative Neurophysiology. Synth. Lect. Biomed. Eng. **3**(1), 1–142 (2008)
16. Troyer, T.W., Miller, K.D.: Physiological gain leads to high isi variability in a simple model of a cortical regular spiking cell. Neural Comput. **9**(5), 971–983 (1997)
17. Van Vreeswijk, C., Sompolinsky, H.: Chaos in neuronal networks with balanced excitatory and inhibitory activity. Science **274**(5293), 1724–1726 (1996)

Efficient Sequential and Parallel Algorithms for Incremental Record Linkage

Abdullah Baihan[1,3], Reda Ammar[1], Robert Aseltine[2], Mohammed Baihan[3], and Sanguthevar Rajasekaran[1(✉)]

[1] Department of Computer Science and Engineering, University of Connecticut, 252 ITEB, 371 Fairfield Way, UConn, Storrs, CT 06269-4155, USA
{abdullah.s.baihan,reda.ammar,sanguthevar.rajasekaran}@uconn.edu
[2] Division of Behavioral Sciences and Community Health and Center for Population Health, UConn Health, 263 Farmington Ave, MC 3910, Farmington, CT 06030-3910, USA
aseltine@uchc.edu
[3] Department of Computer Science, Community College, King Saud University, Riyadh, Saudi Arabia
{aalbaihan,malbaihan}@ksu.edu.sa

Abstract. Given a collection of records, the problem of record linkage is to cluster them such that each cluster contains all the records of one and only one individual. Existing algorithms for this important problem have large run times especially when the number of records is large. Often, a small number of new records have to be linked with a large number of existing records. Linking the old and new records together might call for large run times. We refer to any algorithm that efficiently links the new records with the existing ones as incremental record linkage (IRL) algorithms and in this paper, we offer novel IRL algorithms. Clustering is the basic approach we employ. Our algorithms use a novel random sampling technique to compute the distance between a new record and any cluster and associate the new record with the cluster with which it has the least distance. The idea is to compute the distance between the new record and only a random subset of the cluster records. We can use a sampling lemma to show that this computation is very accurate. We have developed both sequential and parallel implementations of our algorithms. They outperform the best-known prior algorithm (called RLA). For example, one of our algorithms takes 71.22 s to link 100,000 records with a database of 1,000,000 records. In comparison, the current best algorithm takes 140.91 s to link 1,100,000 records. We achieve a very nearly linear speedup in parallel. E.g., we obtain a speedup of 28.28 with 32 cores. To the best of our knowledge, we are the first to propose parallel IRL algorithms. Our algorithms offer state-of-the-art solutions to the IRL problem.

Keywords: Incremental record linkage · Edit distance · Blocking · K-mers · Parallel computing · Hierarchical clustering

1 Introduction

Record linkage is the problem of integrating records from different sources and clustering them so that each cluster has all the records belonging to one and only one individual.

© Springer Nature Switzerland AG 2020
I. Măndoiu et al. (Eds.): ICCABS 2019, LNBI 12029, pp. 26–38, 2020.
https://doi.org/10.1007/978-3-030-46165-2_3

For instance, in the domain of health care, billions of records are stored and maintained in different data sources electronically wherein there could be multiple records for individuals. Record linkage has crucial benefits including cost saving, in analyzing and evaluating disease evolution, in the identification of disease origin and diversity, etc. Record linkage is a challenging problem since, typically, there may not be any global identifier. There are many algorithms to deal with the record linkage problem [1–3]. A naive algorithm for this problem will compare every two records to find the matched records. However, this method might demand too much time. Also, most of the current efficient algorithms [4–8] link only two datasets at a time, while the algorithms in [9–12] link more than two datasets simultaneously. Clustering techniques are typically used for linking records. Clustering is an unsupervised method to divide records into groups, called clusters, so that similar records are classified into the same cluster. In the literature, there are many clustering techniques such as Fuzzy Clustering, Hierarchical Clustering, Artificial Neural Networks for Clustering, Nearest Neighbor Clustering, etc. [27–29].

Very often, to make decisions on an individual (or a small group of individuals), the new records will have to be linked with an already existing (large) database of records. The obvious way of linking the new records will be to add them to the database and perform a linkage on all of the records together. Clearly, this may take an unacceptable amount of time. For instance, consider the case when a physician is in the process of figuring out the right treatment for an ER patient and wants to get information about the patient and other patients with a similar set of symptoms from their previous records. Linking all the records may not be a viable solution. We need algorithms that can link the new records with the old ones efficiently. We refer to such algorithms as incremental record linkage algorithms. Unfortunately, the existing algorithms [13–15] have large run times.

In this paper we propose efficient sequential and parallel IRL algorithms. Our algorithms employ agglomerative hierarchical clustering with single linkage. Also, we have extensively tested our algorithms on a large number of synthetic and real datasets. The results show that our proposed IRLA3 algorithm outperforms the best-known (traditional) record linkage algorithm RLA with a 94.84% consolidated accuracy and takes 137.46 s compared with 164.46 s and 94.12% accuracy for the RLA when the number of new records is 20% of the total number of records in the existing real database. The parallel algorithms achieve a very nearly linear speedup (e.g., the speedup is 28.28 for 32 cores in a single node).

2 Background

Hierarchical Clustering is a common technique used for record linkage [19–22]. Parallel solutions for record linkage are necessary due to performance degradation with increasing data size [9, 10, 23–26]. The Hierarchical Clustering works in two different ways based on how the distance between two clusters is measured: Agglomerative (bottom-up) or Divisive (top-down). The distance between two clusters can be defined in a number of ways. Two popular methods in use are complete linkage and single linkage. If X and Y are any two clusters, then the single linkage distance between them is defined as $\min_{x \in X, y \in Y} d(x, y)$, where $d(x, y)$ is the distance between the records x and y. The complete linkage distance between X and Y is defined as $\max_{x \in X, y \in Y} d(x, y)$ [29].

There are many ways to measure the distance between two records, such as reversal distance, truncation distance, and edit distance. It helps to think of each record as a string (obtained by concatenating its attributes). The RLA in [12] used edit distance which is widely used to measure the differences between records [30]. The edit distance computes the minimum number of edit operations (deletions, substitutions, and insertions) in order to convert one string to another one. For example, if S1 = "cemuterss" and S2 = "computers", in order to convert S1 to S2, the following operations can be used: replace 'e' with 'o' at index 1 of S1, insert 'p' at index 3 of S1, and delete 's' at index 8 of S1.

Even though clustering algorithms are quite useful in record linkage, they typically have large run times. For instance, the hierarchical clustering algorithm takes quadratic time. To speed the clustering algorithms, one of the techniques widely used is blocking. In the literature [31–33], there are many blocking methods such as string-map-based indexing, suffix array-based indexing, sorted neighborhood indexing, canopy clustering, and Q-gram-based indexing. Blocking can be thought of as a first level of coarse clustering. We first coarsely cluster the records into blocks and then perform a fine clustering within the blocks.

Two records will be found in the same block if they share at least one l-mer. There will be a block corresponding to every possible l-mer. The intuition behind blocking is that if two records are very similar then they should share at least one l-mer. The coarse clustering is done using this fact. For example, assume that we block using the LastName attribute. Also let the value of l be 2. There will be a block corresponding every string of length 2. Assume that we have three records with the following LastName values: art, mark, and bill. The 2-mers of art are: ar and rt. The record with art as the LastName will be placed in two blocks one corresponding to ar and the other corresponding to rt. The record of mark will be placed in blocks corresponding to ma, ar, and rk. The record of bill will be placed in three blocks. As we can see, both art and mark will be placed in the same block corresponding to ar. In general, if two records share an l-mer, then they will be placed in the same block corresponding to this l-mer. Nonidentical records might be placed in the same block. The fact that they do not correspond to the same individual will be revealed in the second fine clustering that we do.

After blocking, we have to perform linking only within the blocks. Even though every record may be placed in multiple blocks, the overall run time will still be less (than if we link all the records together).

In the RLA algorithm of [12] records are blocked based on l-mers of one of the attributes (last name, for example). If we assume that the alphabet size is 26, there will be 26^l blocks. Let d be the length of the attribute used for blocking. Then, each record will be found in $(d - l + 1) \le d$ blocks. If n is the number of records in D, then the expected number of records in each block will be $\le \frac{d.n}{26^l}$. After blocking, the blocks will be clustered independently. Finally, linking across the blocks will be done as needed.

One of the currently best known IRL algorithms is in [13]. We refer to this algorithm as GDS. They show that their algorithm performs better than prior algorithms [14, 15]. Since the code of [13] is not accessible, we could not empirically compare our algorithm with those in [13]. The time complexities of the algorithms in [13] are very high. For example, the run time of their GreedyCorr algorithm is $O(|G + \Delta G|^6)$, where G is the original graph and ΔG refers to the subgraph introduced by the new records. In comparison, the run time of our algorithm is only $O(|G|^2)$.

3   Methods

3.1   Our Approaches

We have come up with novel sequential and parallel algorithms for IRL (called IRLA and PIRLA, respectively) that are asymptotically much faster than those of [13]. A summary of our approach follows: Let D be the existing database and let X be a set of new records that have to be incorporated into D. In IRLA we always keep the blocks of D. Given a new record R_{new}, we also block this record based on the same attribute used for blocking D. The record R_{new} will belong to $\leq d$ blocks. We compare R_{new} with each cluster in each of these blocks. As a result, we identify the cluster c with which R_{new} has the least distance. We associate R_{new} with c. To compute the distance between R_{new} and any cluster we use a novel random sampling technique.

The run time of our sequential algorithm is $O(N\frac{d \cdot n}{2^{6l}})$, where n is the number of already existing records, N is the number of new records, and l is the blocking length. The time complexity of our algorithm is $O(|G|^2)$, which is much better than those of [13]. Depending on the sampling technique used to compute the distance between R_{new} and the relevant clusters, there are three variants of IRLA called IRLA1, IRLA2, and IRLA3. In IRLA1, R_{new} will be compared with each record in each cluster. In IRLA2, R_{new} will be compared with a 50% random sample of records from each relevant cluster. In IRLA3, R_{new} will be compared with only one randomly chosen record from each cluster.

3.2   Sequential Algorithms

The IRLA, as shown in Algorithm 1, takes as input four parameters. Note that these parameters (except D_{new}) are outputs of the RLA algorithm [12]. *blockArr* stores the results of blocking the existing database. *Threshold* is a user-specified parameter for identifying records belonging to the same entity (or person). The Set of Single Linkage Clusters (*SLC*) describes the clusters present in the existing database. These clusters would have been obtained using any record linkage algorithm and some blocking technique. In our implementations, we have used the RLA algorithm [12] for generating *SLC* and *blockArr*. The IRLA algorithm processes each new record R_{new} in D_{new} as follows. First, in lines 2–4 of Algorithm 1, our algorithm marks R_{new} as an "*existing*" or an "*unseen*" record. Then, each R_{new} will be added into an existing cluster from *SLC*, only if its type is "*existing*" and the distance (i.e., edit distance) between R_{new} and an existing record is less than the Threshold. Otherwise, it will be added to a new cluster. See lines 6–21 of Algorithm 1. Note that, to create a new cluster, Algorithm 1 utilizes the createNewCluster method that creates a new cluster and adds the record R_{new}, that the method receives, into the new cluster and updates *blockArr* such that the id of the new cluster is added to all the blocks of R_{new}. Finally, in line 23, the algorithm returns an updated *blockArr* and *SLC'* which is a set of all the clusters in *SLC* (updated or not updated) and the newly created clusters.

Algorithm 1 : IRLA (Incremental Record Linkage Algorithm).

Input: a New dataset D_{new}, a Blockage Array (*blockArr*), a Threshold, and a Set of Single Linkage Clusters (*SLC*).
Output: An updated set of Single Linkage Clusters (*SLC'*) and an updated *blockArr*.

```
1.    For each new record R_new in D_new do
2.        R_new type = "existing".
3.        Find all the blocks (BL_new) of the "last name" attribute of R_new.
4.    If any of the blocks in BL_new has not been seen before in ( blockArr) Then R_new type = "unseen".
5.        If R_new type = "unseen" Then
6.            createNewCluster(R_new).
7.        Else
8.            For each block b in BL_new do
9.                Using blockArr, find the set of clusters CL in SLC that correspond to the block b.
10.               For each cluster c in CL do
11.                   Pick a set Q of Ω records from c and add the pair tr=(r, ci), for every r in Q to
                          a Target Record List (TRL).
12.               End For
13.           End For
14.           Compute the edit distance between each r and R_new.
15.           Find the record r that has the minimum edit distance (MED) with R_new.
16.           If (Threshold – MED) >=0
17.               Add R_new to the cluster tr.c
18.           Else
19.               createNewCluster(R_new)
20.           End if
21.       End if
22.   End For
23.   Return the updated set of Clusters (SLC') and an updated blockArr.
```

The expected complexity of the IRLA is $O(NBC\Omega)$, where N is the number of new records in D_{new}, B is the expected number of blocks a new record is placed in, C is the expected number of clusters in each block, and Ω is the expected number of records in each cluster. Note that the complexity of the createNewCluster method is $O(1)$ since it does not depend on the input data (i.e., D_{new}).

3.3 Parallel Algorithm

The PIRLA algorithm, as shown in Algorithm 2, receives five parameters. Note that these parameters (excepting D_{new}) are outputs of the PRLA algorithm [12]. PIRLA processes each new record R_{new} *in* D_{new} as follows. One of the available processors is called the master and the others are called slaves. The master processor performs two steps as shown in lines 1 and 2 of Algorithm 2. In lines 4–7 of Algorithm 2, each processor j marks R_{new} as an "*existing*" or an "*unseen*" record. Then, each R_{new} (in each processor) will be either added into the Existing Clusters List (*ECL*), only if its type is "*existing*" and the distance (i.e., edit distance) between R_{new} and an existing record is less than the Threshold, or into the New Clusters List (*NCL*) otherwise (see lines 8–24 of Algorithm 2). Now, processor j has generated two lists, i.e., NCL and ECL, based on the set of records that j has received. Then, in lines 26–27 of Algorithm 2, the master generates a Master New Clusters List (*MNCL*) and a Master Existing Clusters List (*MECL*), respectively. After that, in lines 28–37 of Algorithm 2, the master iterates through each R in $MNCL$ to add R into an existing cluster of SLC or into a new cluster, and adds all the records in $MECL$ into existing clusters of SLC as shown in lines 38–40 of Algorithm 2. Finally, the algorithm returns the updated set SLC' (see line 41 of Algorithm 2).

The expected time complexity of the PIRLA algorithm is $O(NBC\Omega/P)$, where N is the number of new records in D_{new}, B is the expected number of blocks that a new record is placed in, C is the expected number of clusters of each block, Ω is the expected number of records in each cluster, and P is the number of processors. Note that the complexity of the createNewCluster method is $O(1)$ since it does not depend on the input data (i.e., D_{new}).

Algorithm 2 : PIRLA (Parallel Incremental Record Linkage Algorithm).

Input: a New dataset (D_{new}), a Master Blocks Array (*blockMasterArr*), a Threshold,
               a Set of Single Linkage Clusters (*SLC*), and   Number of slave processors P.
Output: An updated set of Single Linkage Clusters (*SLC'*) and an updated *blockMasterArr*.

1. The Master reads all records from dataset (D_{new}).
2. The Master distributes the records nearly equally among the slave processors.
      // Let D_{new}^{j} be the portion of new data that processor j receives, for $1 \leq j \leq P$.
3. **For** each processor **do**
4. **For** each new record R_{new} in D_{new}^{j} **do**
5. R_{new} type = "existing".
6.   Find the set BLj of all the blocks of the "last name" attribute of R_{new}.
7.   **If** any of the blocks in BLj has not been seen before in (blockMasterArr) **Then**
              R_{new} type = "unseen".
8.   **If** R_{new} type = "unseen" **Then**
9.   Add R_{new} to a New Cluster List (*NCL*)
10.   **Else**
11.   **For** each block b in BLj **do**
12.   Using blockMasterArr, find the set CLj of all the clusters in SLC that
                      correspond to block b.
13.   **For** each cluster c in CLj **do**
14.   Pick a set Q of Ω records from c and add the pair $tr=(r, c)$ to
                      a Target Record List (TRL), for every r in Q.
15.   **End For**
16.   **End For**
17.   Compute the edit distance between each r and R_{new}.
18.   Find the tr that has the minimum edit distance (*MED*) with R_{new}.
19.   **If** (Threshold – MED) ≥ 0
20.   Add R_{new} to an Existing Cluster List (*ECL*)
21.   **Else**
22.   Add R_{new} to a New Cluster List (*NCL*)
23.   **End if**
24.   **End if**
25. **End For**
26. The Master aggregates all *NCL*s from each processor into a Master New Cluster List (*MNCL*);
27. The Master aggregates all *ECL*s from each processor into a Master Existing Cluster List(*MECL*)
      // The Master performs the following two loops
28.   Define a Created Cluster List (*CCL*) to be empty;
29.   **For** each R in *MNCL* **do**
30.   Compute the edit distance (ED) between each C in *CCL* and R.
31.   **If** (Threshold –ED) $>=0$
32.   Add R to the cluster $C.c$
33.   **Else If** (Threshold –ED) <0 for all c
34.   createNewCluster(R)
35.   add the pair $nc(R,C.c)$ to CCL
36.   **End if**
37.   **End For**
38.   **For** each tr in *MECL* **do**
39.   Add R to the cluster $tr.c$
40.   **End For**
41. **Return** the updated set of Clusters (*SLC'*) and an updated *blockMasterArr*.

---

4   Experimental Environments

We have performed all the experiments using real data. We have implemented the sequential versions of all the algorithms in C++ whereas the parallel algorithms have been implemented in C++ with MPI library. For real datasets, due to security and confidentiality reasons, we have deployed all the algorithms on laz-id1 server, UConn Health, Farmington, University of Connecticut. This server has one node and 32 AMD Opteron 6274 @ 2.20 GH cores, 64 GB of RAM, and 1.0 PB of local storage.

The random sampling reduces the run time nicely without much degradation in accuracy. We can explain this with a probabilistic analysis. Let X be a given set of n elements (real numbers for example) and let S be a random sample of X with s elements. Let the minimum element in S be m and let M be the minimum element of X. Then the expected value of m is M. Moreover, we can use Chernoff bounds to show that the value of m will be close to M with a good probability. Consider an element q of S whose rank in S is j. Let the rank of q in X be r_j. Then a lemma of Rajasekaran and Reif [35] states the following:

$$Prob.\left[\left|r_j - j\frac{n}{s}\right| > \sqrt{4\alpha}\frac{n}{\sqrt{s}}\sqrt{\log n}\right] < n^{-\alpha} \text{ for any } \alpha > 0.$$

In our case $j = 1$ and if we apply the above lemma we realize that the difference between the sample minimum and the actual minimum will be small with a high probability.

5 The Datasets

We have run all the experiments on real datasets. We have employed a database called ChimeData that contains actual patient hospitalization data. This database is compiled by the Connecticut Hospital Association (CHA) [34] and maintained by the Connecticut Department of Public Health. The ChimeData has records pertinent to all acute care hospital patients in the state. The dataset from ChimeData that we have used is called "CHIME-partial" that contains information about one of the hospitals in Connecticut. The CHIME-partial contains in total around 1,311,740 patients' records for the years 2012 through 2017. Each record has six attributes: PatientID "PID", LastName, First-Name, gender, date of birth, and race. Our sequential and parallel algorithms have been tested with various numbers of records from CHIME-partial dataset. We have run each algorithm on 1,000,000 records (with an equal number of records from the 6 sub-datasets of CHIME-partial) considered as the existing data. We have also tested the algorithms on datasets of size ranging from 1,000, to 200,000 records taken from CHIME-partial and treated as the new records.

In all the experiments, we have used the LastName attribute as the blocking field and used a threshold value of 2. Blocking has been done on 3-mers. Also, in order to define the distance between any two records, we have employed the edit distance method on five attributes: namely LastName, FirstName, gender, date of birth, and race. We have measured the efficiency and accuracy for each algorithm. For efficiency, we have repeated each experiment 10 times and report the average running times. To measure the accuracy for the incremental algorithms, we have used the PID attribute as the gold standard and computed the accuracy as in the following equation:

$$Accuracy = \frac{the\ total\ number\ of\ records\ with\ correct\ labels}{the\ total\ number\ of\ records} * 100$$

The RLA algorithm does not utilize any prior information. It starts from scratch and links the records. On the other hand, the algorithms IRLA1, IRLA2, and IRLA3 use prior information about known clusters. For each individual, the prior data in general has multiple records. When a new record is given, this record is compared with the existing clusters. If each existing cluster is sufficiently large, then the probability that the new record goes to an incorrect cluster is low. As a result, the accuracies of IRLA1, IRLA2, and IRLA3 (on the new records) are better than that of RLA (on all the records). It may not be fair if we compare the accuracy of the new algorithms (as measured only on the new records) with the accuracy of RLA (as measured on all the records). To make a fair comparison we define the following consolidated accuracy:

$$Consolidated\ Accuracy = \frac{N_{old} \times A_{old} + N_{new} \times A_{new}}{N_{old} + N_{new}},$$

where N_{old} is the number of existing records, N_{new} is the number of new records, A_{old} is the accuracy obtained while linking the existing records (using RLA, for example), and A_{new} is the accuracy obtained by the new incremental algorithms for linking the new records.

6 Results

We have performed many experiments to determine the effectiveness of our proposed incremental record linkage algorithms IRLAs. We have tested IRLA1, IRLA2, and IRLA3 and compared their performance with that of the best-known RLA [12] in terms of the running time and the accuracy. These comparisons were done to understand for what sizes of the new dataset D_{new} will the algorithms IRLA1, IRLA2, and IRLA3 be faster than running RLA on the old and new datasets together. In the next sections we report the results for sequential and parallel algorithms on real datasets.

6.1 Results on Real Datasets for Sequential Algorithms

In this section, we present the implementation results of our proposed sequential algorithms, namely, IRLA1, IRLA2, and IRLA3 and the best-known sequential RLA [12] on real datasets. We assume that we already have a dataset with 1,000,000 records from CHIME-partial. And we test the new algorithms when the number of new records ranges from 1,000 to 200,000.

Table 1 summarizes the comparison of our proposed sequential IRLA algorithms with the RLA algorithm. The first column of Table 1 shows the number of records in the existing dataset. The second column lists the number of new records. For example, if the number of new records is 1,000 it means that the algorithm RLA has been used to link 1,001,000 records (i.e., with 1,000,000 records in the existing dataset and 1,000 new records).

IRLA algorithms have been run on 1,000 new records only (to be linked with an existing database of size 1,000,000). The running times shown for RLA are the total times for linking the old and new records while the running times shown for IRLA1, IRLA2, and IRLA3 are for linking the new records only. From Table 1, we clearly see that IRLA1, IRLA2 and IRLA3 outperform RLA in terms of the running time and the accuracy when the number of new records is 3.85%, 7.71%, and 20% respectively, of the

Table 1. Comparison results of our three proposed sequential IRLA algorithms with the best-known sequential RLA algorithm on real datasets.

Size of the existing dataset	Size of the new dataset	RLA		IRLA1		IRLA2		IRLA3	
		Time	Accuracy	Time	Consolidated accuracy	Time	Consolidated accuracy	Time	Consolidated accuracy
1,000,000	1,000	114.33	93.60%	4.31	93.606%	2.47	93.605%	1.29	93.605%
1,000,000	5,000	116.17	93.60%	20.37	93.630%	12.15	93.627%	5.11	93.624%
1,000,000	10,000	118.59	93.61%	38.91	93.670%	24.42	93.664%	10.17	93.657%
1,000,000	25,000	121.26	93.66%	91.08	93.808%	56.13	93.788%	23.79	93.777%
1,000,000	50,000	126.78	93.72%	160.89	94.011%	97.92	93.970%	42.18	93.932%
1,000,000	75,000	132.6	93.78%	215.97	94.206%	131.13	94.153%	58.05	94.093%
1,000,000	100,000	140.91	93.84%	253.41	94.390%	158.79	94.325%	71.22	94.245%
1,000,000	125,000	145.41	93.92%	336.3	94.584%	205.98	94.493%	90.84	94.411%
1,000,000	150,000	151.89	93.99%	398.4	94.762%	238.41	94.662%	107.76	94.564%
1,000,000	175,000	158.04	94.06%	451.23	94.933%	268.59	94.820%	121.44	94.709%
1,000,000	200,000	164.46	94.12%	503.88	95.090%	304.56	94.965%	137.46	94.848%

total number of records in the existing real database. For every new record, the IRLA1 algorithm will compare this new record with all the records in each cluster, and IRLA2 will compare each new record with a random half of the records in each cluster. As a result, the run times of these algorithms increase linearly with the number of new records. For this reason, when the number of records is very large, there may not be any gain in using an incremental linkage algorithm. Besides, on the average, IRLA3 is 2.25 times faster than IRLA2; IRLA2 is 1.65 times faster than IRLA1; and IRLA3 is 3.71 times faster than IRLA1.

Figure 1 displays a runtime comparison of our IRLA algorithms with the RLA algorithm on 1,001,000 to 1,200,000 records of real datasets. Also, it shows that the break-even points for IRLA1 and IRLA2 are 38,567 and 77,115 records, respectively. As a result, our proposed IRLA3 is still outperforming the best-known RLA [12] with 94.84% consolidated accuracy and takes 137.46 s compared with 164.46 s and 94.12% accuracy for the RLA on 200,000 incremental records.

6.2 Results on Real Datasets for Parallel Algorithms

Table 2 summarizes the comparison results in terms of running time in seconds, speedup, and the accuracy of our three proposed PIRLA algorithms with the best-known PRLA algorithm [12]. All the algorithms have been run on 1,200,000 records of a real dataset on 1 to 32 cores. Figure 2 shows that the speedup for our proposed PIRLA algorithms is almost linear. For instance, the speedup for PIRLA3 is 7.52 for 8 cores, 14.77 for 16 cores, and 28.28 for 32 cores in a single node. Also, as Table 2 shows, the consolidated accuracy of our PIRLA algorithms is more than 94.77% compared to 94.04% for the PRLA algorithm.

Table 2. Comparison results of our three proposed parallel IRLA algorithms with the best-known parallel RLA algorithm on real datasets on multiple cores.

No of Cores	PRLA			PIRLA1			PIRLA2			PIRLA3		
	Time	Speedup	Accuracy	Time	Speedup	Consolidated accuracy	Time	Speedup	Consolidated accuracy	Time	Speedup	Consolidated accuracy
1	164.46	1.00	94.12%	503.88	1.00	95.090%	304.56	1.00	94.965%	137.46	1.00	94.848%
2	86.55	1.90	94.10%	263.81	1.91	95.072%	158.62	1.92	94.947%	71.22	1.93	94.828%
4	44.09	3.73	94.09%	134.36	3.75	95.060%	81.11	3.76	94.935%	36.36	3.78	94.818%
8	22.16	7.42	94.07%	67.72	7.44	95.040%	40.77	7.47	94.915%	18.27	7.52	94.800%
16	11.24	14.63	94.06%	34.32	14.68	95.030%	20.67	14.73	94.905%	9.31	14.77	94.788%
32	5.85	28.11	94.04%	17.89	28.15	95.012%	10.79	28.21	94.885%	4.86	28.28	94.770%

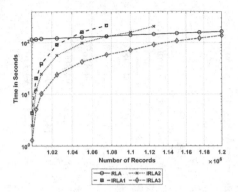

Fig. 1. Runtime comparison of our three proposed sequential IRLA algorithms with the best-known sequential RLA algorithm on real datasets. (x-axis denotes to total records in millions; y-axis corresponds to the time in seconds).

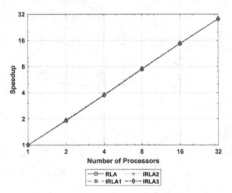

Fig. 2. The speedup of our three proposed parallel IRLA algorithms with the best-known parallel RLA algorithm on real datasets. (x-axis corresponds to the number of processors; y-axis denotes speed up).

7 Conclusions

We have presented novel algorithms for the crucial problem of incremental linkage. Our algorithms employ a novel random sampling technique and single linkage clustering and outperform previous best-known algorithms in this category. Our proposed IRLA3 algorithm outperforms the best-known RLA algorithm when the number of incremental records is up to around 20% of the total number of existing records. To the best of our knowledge, we are the first who have proposed a parallel incremental record linkage algorithm. Our sequential and parallel algorithms have been tested on synthetic and real datasets. They achieve a consolidated accuracy of up to 95.09%, take less run times, and attain a very nearly linear speedup compared to the best-known algorithms in this category. Our parallel algorithms can be applied on hundreds of processors over millions of records. Therefore, our proposed algorithms, IRLA and PIRLA offer state-of-the-art solutions for the incremental record linkage problem.

Acknowledgment. This work has been supported in part by the following NSF grants: 1447711, 1514357, 1743418, and 1843025. Also, this project was supported by King Saud University, Deanship of Scientific Research, Community College Research Unit. Data for this study were obtained from the Connecticut Department of Public Health. The Connecticut Department of Public Health does not endorse or assume any responsibility for any analyses, interpretations or conclusions based on the data. The Human Investigations Committee of the Department of Public Health approved this study.

References

1. Koudas, N., Sarawagi, S., Srivastava, D.: Record linkage: similarity measures and algorithms. In: Proceedings of the 2006 ACM SIGMOD International Conference on Management of data, pp. 802–803. ACM (2006)
2. Gomatam, S., Carter, R., Ariet, M., et al.: An empirical comparison of record linkage procedures. Stat. Med. **21**(10), 1485–1496 (2002). https://doi.org/10.1002/sim.1147. PMID: 12185898
3. Winkler, W.E.: Overview of record linkage and current research directions. In: Bureau of the Census. Citeseer (2006)
4. Christen, P., Churches, T., Hegland, M.: Febrl – a parallel open source data linkage system. In: Dai, H., Srikant, R., Zhang, C. (eds.) PAKDD 2004. LNCS (LNAI), vol. 3056, pp. 638–647. Springer, Heidelberg (2004). https://doi.org/10.1007/978-3-540-24775-3_75
5. Christen, P.: Febrl—a freely available record linkage system with a graphical user interface. In: Second Australasian Workshop on Health Data and Knowledge Management, vol. 80, pp. 17–25 (2008)
6. Jurczyk, P., Lu, J.J., Xiong, L., et al.: FRIL: a tool for comparative record linkage. In: AMIA Annual Symposium Proceedings, vol. 2008. American Medical Informatics Association, p. 440 (2008)
7. Jurczyk, P., Lu, J.J., Xiong, L., et al.: Fine-grained record integration and linkage tool. Birth Defects Res. Part A: Clin. Mol. Teratol. **82**(11), 822–829 (2008). https://doi.org/10.1002/bdra.2052
8. Lee, M.L., Ling, T.W., Low, W.L.: IntelliClean: a knowledge-based intelligent data cleaner. In: Proceedings of the Sixth ACM SIGKDD International Conference on Knowledge Discovery and Data Mining, pp. 290–294. ACM (2000)
9. Mi, T., Rajasekaran, S., Aseltine, R.: Efficient algorithms for fast integration on large data sets from multiple sources. BMC Med. Inform. Decis. Making **12**(1), 59 (2012). https://doi.org/10.1186/1472-6947-12-59. PMID: 22741525
10. Mi, T., Aseltine, R., Rajasekaran, S.: Data integration on multiple data sets. In: 2008 IEEE International Conference on Bioinformatics and Biomedicine. BIBM 2008, pp. 443–446. IEEE (2008)
11. Li, X., Shen, C.: Linkage of patient records from disparate sources. Stat. Methods Med. Res. **22**(1), 31–38 (2013). https://doi.org/10.1177/0962280211403600. PMID: 21665896
12. Mamun, A.A., Mi, T., Aseltine, R., Rajasekaran, S.: Efficient sequential and parallel algorithms for record linkage. J. Am. Med. Inform. Assoc. **21**(2), 252–262 (2014). https://doi.org/10.1136/amiajnl-2013-002034. PMID: 24154837
13. Gruenheid, A., Dong, X.L., Srivastava, D.: Incremental record linkage. Proc. VLDB Endow. **7**(9), 697–708 (2014)
14. Whang, S.E., Garcia-Molina, H.: Entity resolution with evolving rules. Proc. VLDB **3**(1), 1326–1337 (2010)

15. Whang, S.E., Garcia-Molina, H.: Incremental entity resolution on rules and data. VLDB J. **23**(1), 77–102 (2013). https://doi.org/10.1007/s00778-013-0315-0
16. Christen, P.: Data Matching: Concepts and Techniques for Record Linkage, Entity Resolution, and Duplicate Detection. Springer, Heidelberg (2012). https://doi.org/10.1007/978-3-642-31164-2
17. Gu, L., Baxter, R., Vickers, D., et al.: Record linkage: current practice and future directions. CSIRO Mathematical and Information Sciences Technical Report 3/83 (2003)
18. Brizan, D.G., Tansel, A.U.A.: Survey of entity resolution and record linkage methodologies. Commun. IIMA **6**(3), 5 (2015)
19. Rajasekaran, S.: Efficient parallel hierarchical clustering algorithms. IEEE Trans. Parallel Distrib. Syst. (6), 497–502 (2005). https://doi.org/10.1109/tpds.2005.72
20. Li, X.: Parallel algorithms for hierarchical clustering and cluster validity. IEEE Trans. Pattern Anal. Mach. Intell. **12**(11), 1088–1092 (1990). https://doi.org/10.1109/34.61708
21. Olson, C.F.: Parallel algorithms for hierarchical clustering. Parallel Comput. **21**(8), 1313–1325 (1995). https://doi.org/10.1016/0167-8191(95)00017-i
22. Wu, C.H., Horng, S.J., Tsai, H.R.: Efficient parallel algorithms for hierarchical clustering on arrays with reconfigurable optical buses. J. Parallel Distrib. Comput. **60**(9), 1137–1153 (2000). https://doi.org/10.1006/jpdc.2000.1644
23. Kawai, H., Garcia-Molina, H., Benjelloun, O., et al.: P-swoosh: parallel algorithm for generic entity resolution (2006)
24. Kim, H.S., Lee, D.: Parallel linkage. In: Proceedings of the Sixteenth ACM Conference on Conference on Information and Knowledge Management, pp. 283–292. ACM (2007)
25. Kirsten, T., Kolb, L., Hartung, M., et al.: Data partitioning for parallel entity matching. arXiv preprint arXiv:10065309 (2010)
26. Bianco, G.D., Galante, R., Heuser, C.A.: A fast approach for parallel deduplication on multicore processors. In: Proceedings of the 2011 ACM Symposium on Applied Computing, pp. 1027–1032. ACM (2011)
27. Steorts, R.C., Ventura, S.L., Sadinle, M., Fienberg, S.E.: A comparison of blocking methods for record linkage. In: Domingo-Ferrer, J. (ed.) PSD 2014. LNCS, vol. 8744, pp. 253–268. Springer, Cham (2014). https://doi.org/10.1007/978-3-319-11257-2_20
28. Christen, P.: A survey of indexing techniques for scalable record linkage and deduplication. IEEE Trans. Knowl. Data Eng. **24**(9), 1537–1555 (2012). https://doi.org/10.1109/TKDE.2011.127
29. Bachteler, T., Reiher, J., Schnell, R.: Similarity filtering with multibit trees for record linkage. German Record Linkage Center, Nuremberg, Working Paper WP-GRLC-2013-02 (2013)
30. Levenshtein, V.I.: Binary codes capable of correcting deletions, insertions, and reversals. In: Soviet Physics Doklady, vol. 10, pp. 707–710 (1966)
31. Jain, A.K., Murty, M.N., Flynn, P.J.: Data clustering: a review. ACM Comput. Surv. (CSUR) **31**(3), 264–323 (1999). https://doi.org/10.1145/331499.331504
32. McCallum, A., Nigam, K., Ungar, L.H.: Efficient clustering of high-dimensional data sets with application to reference matching. In: Proceedings of the Sixth ACM SIGKDD International Conference on Knowledge Discovery and Data Mining, pp. 169–178. ACM (2000)
33. Rokach, L., Maimon, O.: Clustering methods. In: Maimon, O., Rokach, L. (eds.) Data Mining and Knowledge Discovery Handbook, pp. 321–352. Springer, Boston (2005). https://doi.org/10.1007/0-387-25465-X_15
34. Connecticut Hospital Association: https://cthosp.org/member-services/chimedata/chimedata-overview/
35. Rajasekaran, S., Reif, J.H.: Derivation of randomized sorting and selection algorithms. In: Paige, R., Reif, J., Watcher, R. (eds.) Parallel Algorithm Derivation and Program Transformation. The Springer International Series in Engineering and Computer Science, vol. 231, pp. 187–205. Springer, Boston (1993). https://doi.org/10.1007/978-0-585-27330-3_6

Autoencoder Based Methods for Diagnosis of Autism Spectrum Disorder

Sakib Mostafa⬤, Wutao Yin⬤, and Fang-Xiang Wu(✉)⬤

University of Saskatchewan, Saskatoon, Canada
sakib.mostafa@usask.ca, {wuy272,faw341}@mail.usask.ca

Abstract. Autism Spectrum Disorder (ASD) is a neurological disorder that affects a person's behavior and social interaction. Integrating machine learning algorithms with neuroimages a diagnosis method can be established to detect ASD subjects from typical control (TC) subjects. In this study, we develop autoencoder based ASD diagnosis methods. Firstly, we design an autoencoder to extract high-level features from raw features, which are defined based on eigenvalues and centralities of functional brain networks constructed with the entire Autism Brain Imaging Data Exchange 1 (ABIDE 1) dataset. Secondly, we use these high-level features to train several traditional machine learning methods (SVM, KNN, and subspace discriminant), which achieve the classification accuracy of 72.6% and the area under the receiving operating characteristic curve (AUC) of 79.0%. We also use these high-level features to train a deep neural network (DNN) which achieves the classification accuracy of 76.2% and the AUC of 79.7%. Thirdly, we combine the pre-trained autoencoder with the DNN to train it, which achieves the classification accuracy of 79.2%, and the AUC of 82.4%. Finally, we also train SVM, KNN, and subspace discriminant with the features extracted from the combination of the pre-trained autoencoder and the DNN which achieves the classification accuracy of 74.6% and the AUC of 78.7%. These results show that our proposed methods for diagnosis of ASD outperform state-of-the-art studies.

Keywords: Autism Spectrum Disorder · Deep learning · Autoencoder · Functional magnetic resonance imaging · Brain networks

1 Introduction

Autism spectrum disorder (ASD) is a neuro dysfunction. It covers a wide range of behavioral abnormality such as impaired social skills, co-occurring behaviors, reduced speech, attention deficit, etc. The term spectrum in ASD refers to different conditions of subjects. ASD affects different subjects differently. According to a report in [1], ASD subjects are 2.5 times more likely to cause premature death than healthy controls (HC). There is no acute treatment for ASD. However, a timely and precise diagnosis can help the family take preliminary and effective steps to ensure the normal life of a patient.

Patients start to show symptoms of ASD during the first three years of life. However, sometimes they grow normally and then start showing symptoms at the age of 18 to 36 months. Despite the extensive research into the diagnosis of ASD, it has been a difficult

© Springer Nature Switzerland AG 2020
I. Măndoiu et al. (Eds.): ICCABS 2019, LNBI 12029, pp. 39–51, 2020.
https://doi.org/10.1007/978-3-030-46165-2_4

task to accomplish. Apart from monitoring the behavior and development of patients, there are no other perceive signs for effective diagnosis of ASD. Traditional diagnosis methods include Autism Diagnostic Observation Schedule [2] and Autism Diagnostic Interview [3], where the diagnosis is done through the observation of behavior and interview of a patient. However, these diagnosis methods are time-consuming and can be false sometimes, as there are no specific behaviors that can be described as ASD. Therefore, it is necessary to invent ways that can diagnose ASD more accurately and more efficiently without relying on pure behavioral patterns.

The study of the neuroimages has been widely used to understand different brain diseases. Machine learning techniques are combined with neuroimages to find biomarkers and establish a diagnosis method for ASD. A better approach for studying the neuroimages is to use the Magnetic Resonance Imaging (MRI) as it can be used to extract information about the structural and functional activities of a brain. Out of the different MRI techniques, the resting-state functional MRI (rs-fMRI) provides information about the neural activities of a person's brain. However, instead of studying the raw MRI images, a more effective way is the graph-theoretic or network based approaches. In network based approaches a brain network is created from MRI images by dividing a brain into different regions of interest (ROIs). The main components of a network are nodes and edges connecting nodes. In a functional brain network, ROIs are nodes while the Pearson Correlation Coefficient (PCC) of the time series measurement of ROIs between a pair of nodes is used to determine their edge weight.

There are a number of studies where the information from brain networks is incorporated into machine learning techniques for the diagnosis of ASD subjects from typical control (TC) subjects. In [4], the authors proposed an ASD diagnosis method using brain images from the Autism Brain Imaging Data Exchange 1 (ABIDE 1) dataset [5]. They pre-trained a deep neural network (DNN) classifier using a stacked denoising autoencoder, where the input to the network was the PCC of all pairwise ROIs. They reported a classification accuracy of 70.0% for all the sites of ABIDE 1 and the average classification accuracy of 52.0% for individual sites. In [6], a classification accuracy of 70.1% was reported for the diagnosis of ASD using the ABIDE 1 dataset. They used an autoencoder to pre-train a single layer perceptron. The autoencoder functioned as a feature extractor and the perceptron has worked as the classifier. They also reported an average classification accuracy of 63.0% for the individual sites of ABIDE 1. The Riemannian geometry of the functional connectivity was studied in [7]. Using the log-Euclidean and affine-invariant Riemannian matrices in the machine learning algorithms, their method achieved an accuracy of 71.1% for the entire ABIDE 1 dataset. A DNN based ASD diagnosis method was presented in [8]. In their study, the authors have extracted features from brain networks and used the F-score to select the dominant features. The features were then used in a DNN classifier, which consisted of two stacked autoencoders connected to a softmax function for classification. They achieved a classification accuracy of 90.4%. However, they used data from only a single site of the ABIDE 1 dataset. In some other studies, higher classification accuracy for diagnosing ASD subjects was also achieved [9, 10]. Those studies also only included some parts of the ABIDE 1 dataset. They haven't proved the generalization of the methods by including the entire ABIDE 1 dataset. According to [11], the studies related to the diagnosis of ASD tended to have high accuracy for the small sample size and the accuracy declined with the increase of the sample size. Therefore, it is appealing to establish a diagnosis method that has a high classification accuracy using the whole ABIDE 1 dataset.

In [12] a set of new features was proposed for the diagnosis of ASD. They proposed the spectrum of the Laplacian matrix of brain networks as raw features and combined them with three network centralities: assortativity, clustering coefficient, and average degree. Using traditional machine learning algorithms, they have achieved a classification accuracy of 77.7%, which is considered to be the highest at that time for the entire ABIDE 1 dataset.

In this study, we are extending the work proposed in [12]. We use the same 871 subjects from the ABIDE 1 dataset. We adopt the raw features proposed in [12] i.e. the spectrum of Laplacian matrices, assortativity, clustering coefficient, and average degree of brain networks to train an autoencoder to extract the high-level features. Then use these high-level features train several machine learning models such as SVM, KNN, subspace discriminant, and DNN to obtain classifiers for the diagnosis of ASD, whose average performances are comparable to those of state-of-the-art methods. We further combine the pre-trained autoencoder with a DNN and train this classifier, which achieves the classification accuracy of 79.2%, and the area under the receiving operating characteristic curve (AUC) of 82.4%, which are better than results reported in [12]. We also train SVM, KNN, and subspace discriminant with the features extracted from the combination of the pre-trained autoencoder and the DNN which achieves the classification accuracy of 74.6% and the AUC of 78.7%, which is better than the studies in [4, 6, 7, 12, 13].

2 Materials and Methods

2.1 ABIDE

ABIDE 1 [14] dataset is a very common dataset for studies related to the diagnosis of AD. In this dataset, there are rs-fMRI images, T1-weighted images, and phenotypic information of subjects suffering from ASD along with TC subjects. The data are collected from 17 different sites. In ABIDE 1 there are data of 1112 subjects. Out of the 1112 subjects, 539 are ASD subjects and 573 are TC subjects. Analyzing the image acquisition techniques and phenotypic information provided in [5], we can say that ABIDE 1 covers a wide range of scanners, scanning parameters, ages, etc. Because of the heterogeneity of the subjects and variance between inter-site data, ABIDE 1 is a very complicated dataset to work with. As a result, a diagnosis method that works well with the ABIDE 1 should be able to tackle the variations in the real-world scenarios. To be consistent with the study [12] and compare our study with other studies [4, 6, 7, 12, 13] we have experimented with the same 871 subjects, among which 403 are ASD subjects and 468 are TC subjects.

2.2 Data Preprocessing and Brain Networks

Before creating brain networks from the rs-fMRI images it is necessary to preprocess images. For the preprocessing, we use the AFNI (Analysis of Functional Neuroimages) [15] and FSL (FMRIB's Software Library) [16] software packages. We use both rs-fMRI and T1-weighted images. However, the only purpose of using T1-weighted images is to register rs-fMRI images to the standard space. We adopt the identical preprocessing steps as in [12] to remove the noises from rs-fMRI data.

A network consists of nodes and edges, where nodes are connected through edges. To define nodes in the brain network, we adopt the 264 ROIs based parcellation scheme [17]. Specifically, we divide the brain into 264 ROIs and defined each ROI as a node of the network. Then we obtain the time-series measurements of each ROI. The PCC between the time-series measurements of each pair of ROI is the edge weights of the network. The PCC, r_{xy} of any two time series, x and y is calculated as follows

$$r_{xy} = \frac{\sum_{b=1}^{s} (x_b - \bar{x})(y_b - \bar{y})}{\sqrt{\sum_{b=1}^{s} (x_b - \bar{x})^2}\sqrt{\sum_{b=1}^{s} (y_b - \bar{y})^2}} \tag{1}$$

where s is the length of time series, x_b and y_b are the b-th component of x and y, respectively, $\bar{x}$ and $\bar{y}$ are the means of x and y, respectively. The PCC ranges from $+1$ to -1, where a positive PCC indicates the similarity between the activation of the ROIs and a negative PCC indicates the dissimilarity between the activation of the ROIs.

2.3 Feature Extraction

To create a matrix representation of a brain network and for the simplicity of feature extraction, we define a 264×264 connectivity matrix for each network. Each row and column of the connectivity matrix represent a node in the network and each element of the matrix represents edge weights. Similar to [12], we apply different threshold values, $T > 0$ to filter noise edges in the connectivity matrix, $CM = (cm_{i,j})_{n \times n}$ and thus to create the adjacency matrix, $A = (a_{i,j})_{n \times n}$ as follows

$$a_{i,j} = \begin{cases} 1, & if \ cm_{i,j} \geq T \\ -1, & if \ cm_{i,j} \leq -T \\ 0, & if \ i = j \\ 0, & otherwise \end{cases} \tag{2}$$

Then we calculate the Laplacian matrix of an undirected graph $G = (V, E)$ from the adjacency matrix as follows

$$L(G) = D(G) - A(G) \tag{3}$$

where $A(G)$ is the adjacency matrix and $D(G)$ is the degree matrix, $D = (d_{i,j})_{n \times n}$, which is calculated as follows

$$d_{i,j} = \begin{cases} \sum_{k=1}^{n} a_{i,k}, & if \ i = j \\ 0, & otherwise \end{cases} \tag{4}$$

Therefore, the Laplacian matrix is the difference between the degree matrix and the adjacency matrix.

After creating the matrix representation of a brain network, we calculate the features i.e. the spectrum of the Laplacian matrix, assortativity, clustering coefficient, and average degree. The spectrum of a matrix is the set of all eigenvalues of that matrix. An eigenvalue, λ of a matrix M can be obtained by solving its characteristic equation

$$P(\lambda) = \det(M - \lambda I) = 0 \tag{5}$$

where I is an identity matrix.

Apart from the spectrum, we also calculate the topological centralities: assortativity, clustering coefficient, and average degree, as the raw features. To calculate the assortativity and the clustering coefficient, the adjacency matrix A is transformed to $\bar{A} = \left(\bar{a}_{i,j}\right)_{n \times n}$ as follows

$$\bar{a}_{i,j} = \begin{cases} 1, & if \ a_{i,j} > 0 \\ 0, & otherwise \end{cases} \tag{6}$$

The assortativity is then calculated using the MATLAB function defined in [18]. To calculate the clustering coefficient, at first, we compute the number of triangles of each node (denoted by β_G) as follows

$$\beta_G = \boldsymbol{diag}\left(\bar{A} \times U\left(\bar{A}\right) \times \bar{A}\right) \tag{7}$$

where $diag$ is the MATLAB function which returns the diagonal elements of a matrix and $U\left(\bar{\mathbf{A}}\right)$ is the upper triangular matrix of $\bar{A}$.

The clustering coefficient, C is calculated as follows

$$C = \frac{1}{f}\left(\sum_{i \in V} 2 \times \left(\frac{\beta_G(i)}{d_i \times (d_i - 1)}\right)\right) \tag{8}$$

where for a network $G = (V, E)$, f is the total number of nodes in the network and d_i is the degree of node i.

The average degree (denoted by Q) of a network is calculated directly from the adjacency matrix $A = \left(a_{i,j}\right)_{n \times n}$ as follows

$$Q = \frac{2}{f} \times \sum_{i=1}^{f} \sum_{j=1}^{f} a_{i,j} \tag{9}$$

where for a network $G = (V, E)$, f is the total number of nodes in the network.

2.4 Feature Extraction

The feature normalization is an important step for machine learning classifiers as the uneven features may bias the results. Therefore, we have normalized the features before applying them to the DNN. We have used MATLAB to calculate the spectrum of the Laplacian matrix. As the size of the Laplacian matrix is 264×264, so there are 264 eigenvalues for each subject. The eigenvalues are sorted in ascending order for each subject and they can range from $-\infty$ to $+\infty$. However, when the eigenvalues (say the first eigenvalue of each subject) are normalized, the contribution of each eigenvalue is

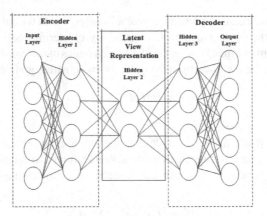

Fig. 1. Example of a simple autoencoder

measured over all the subjects. The contribution of that eigenvalue for that subject is ignored in this scenario. So, in this study, we have normalized the eigenvalues for a subject rather than normalizing each eigenvalue over all subjects using the equation as follows

$$z' = \frac{z - \min(z)}{\max(z) - \min(z)} \tag{10}$$

where z is an original value, and z' is the normalized value. After this normalization, the maximum value and the minimum value in the spectrum of each Laplacian matrix are 1 and 0, respectively, which are excluded from the raw feature set.

We haven't applied normalization to the assortativity and clustering coefficient, as those are normalized when calculated. However, the average degree of the network is normalized over all the subjects using Eq. 10.

2.5 Proposed Autoencoder

In simple words, an autoencoder is a neural network that learns to replicate its input. Figure 1 shows the architecture of our proposed autoencoder. There are three main components of the autoencoder: encoder, latent space representation, and decoder. The encoder compresses the input data and creates a latent space representation (high-level feature), which is the compressed representation of the input data, and the decoder tries to reconstruct the input data from the latent space representation. There is a total of three hidden layers, one input layer, and one output layer in our proposed autoencoder. In the consecutive layers of the encoder, we have decreased the number of neurons from 267 (input layer) to 200 (hidden layer 1). Then the information learned by the encoder is projected into the latent space representation through 10 neurons (hidden layer 2). Then in the decoder, we have increased the neurons from 200 (hidden layer 3) to 267 (output layer). The decoder reconstructs the data from the latent view representation. Therefore, the autoencoder is trying to recreate x' from the input data x by minimizing the reconstruction error, $L(x, x')$, where L is the mean squared error between original input and the consequent reconstruction. Rather than using the autoencoder to copy the input information to the output, the latent space is used to teach the network the useful attributes of the data.

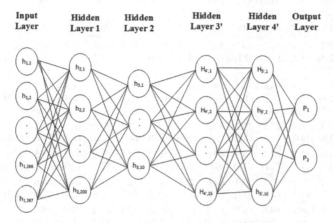

Fig. 2. Proposed deep neural network classifier

The autoencoder can be used both as a feature extractor and a classifier for the diagnosis of ASD. Using the autoencoder as a classifier, the information learned from the autoencoder is incorporated into a classifier model for the classification [8]. Deploying the autoencoder as a feature extractor, the latent view representation of the data is considered to be the high-level features, and these features are then used for diagnosis [19]. In the case of a feature extractor, the autoencoder is popular because it can approximate the nonlinear relation of features through the nonlinear activation functions. Filtering, wrapping, and embedding are some of the common feature selection methods. However, all these methods can only find the linear relation among the features. The embedded method can be used to find the nonlinear relationship of the features by using a nonlinear kernel, but the learning of the model greatly depends on the kernel [20]. Also, in [20] the authors have shown that an autoencoder based feature selection scheme works better than the traditional feature selection algorithms.

3   Results and Discussion

At first, we implement an autoencoder based feature extraction method. As mentioned, the decoder reconstructs the input data in the output from the latent space representation. Therefore, if the reconstruction error is small enough, it means the latent space have learned the salient features of the input data. Thus, the latent space can produce discriminate and salient representation (high-level features) of the input data. These features can be used in machine learning algorithms for the purpose of classification between ASD subjects and TC subjects. To evaluate the performance of the autoencoder based feature extractor, we divide the entire dataset into 80% training data (697 subjects) and 20% testing data (174 subjects). Then we train the autoencoder using only the training

Table 1. Performance analysis of the autoencoder based feature selector

Thresholding condition	Linear SVM (%) (stdv)		Medium Gaussian SVM (%) (stdv)		Coarse Gaussian SVM (%) (stdv)		Medium KNN (%) (stdv)		Cosine KNN (%) (stdv)		Weighted KNN (%) (stdv)		Subspace discriminant (%) (stdv)	
	ACC	AUC	ACC	AUC	ACC	AUC	ACC	AUC	ACC	AUC	ACC	AUC	ACC	AUC
All edges with positive PCC	57.2 (2.0)	57.4 (3.7)	62.2 (1.8)	70.2 (2.1)	54.7 (1.9)	57.5 (1.5)	64.9 (3.2)	70.5 (3.0)	64.8 (1.2)	70.8 (2.0)	68.1 (3.1)	74.8 (2.5)	59.9 (1.4)	58.2 (2.0)
All edges with negative PCC	61.2 (2.1)	65.3 (1.9)	64.6 (1.7)	71.3 (1.2)	59.7 (2.7)	62.0 (2.5)	65.4 (1.3)	70.5 (1.1)	65.8 (3.1)	70.4 (2.0)	65.5 (1.7)	73.1 (1.3)	61.3 (2.4)	65.7 (1.8)
T = 0	61.3 (3.6)	66.4 (4.0)	67.5 (3.0)	72.2 (1.6)	57.3 (5.4)	62.2 (5.3)	63.6 (2.7)	68.7 (4.6)	63.3 (2.8)	68.1 (4.3)	65.3 (1.7)	71.0 (3.4)	61.0 (3.0)	64.6 (3.3)
T = 0.1	63.4 (2.1)	67.2 (1.1)	67.4 (2.1)	70.9 (0.9)	59.6 (0.6)	64.7 (0.9)	65.2 (1.6)	71.8 (1.8)	67.2 (2.1)	74.1 (1.0)	65.5 (1.6)	73.4 (1.6)	59.9 (1.5)	65.6 (1.7)
T = 0.2	66.7 (1.7)	71.0 (1.2)	69.3 (2.2)	73.3 (0.7)	67.6 (3.2)	71.6 (3.0)	66.1 (1.8)	72.9 (1.2)	68.5 (2.5)	74.7 (2.9)	66.6 (2.5)	74.3 (1.3)	68.9 (2.6)	72.2 (1.6)
T = 0.3	63.8 (1.1)	69.1 (0.7)	64.6 (2.9)	69.5 (2.6)	61.2 (3.8)	63.8 (3.2)	66.2 (3.0)	72.2 (2.6)	65.2 (2.2)	71.3 (1.8)	65.4 (2.9)	72.5 (3.2)	65.2 (2.9)	67.3 (1.6)
T = 0.4	**68.0 (1.2)**	**71.4 (1.8)**	**72.2 (0.7)**	**78.0 (1.1)**	**66.9 (3.6)**	**70.9 (3.1)**	**72.4 (2.1)**	**77.7 (1.9)**	**72.6 (1.3)**	**79.0 (1.1)**	**72.4 (2.1)**	**77.6 (1.4)**	**69.6 (2.6)**	**72.1 (1.7)**
T = 0.5	58.9 (5.6)	61.3 (8.0)	69.9 (2.1)	73.1 (1.5)	55.5 (3.5)	61.8 (2.8)	65.3 (1.8)	70.9 (1.4)	64.7 (2.2)	69.1 (2.3)	69.3 (1.7)	77.2 (2.0)	60.1 (2.3)	61.9 (4.8)
T = 0.6	54.3 (1.6)	55.1 (2.3)	62.9 (2.3)	62.6 (4.6)	56.0 (2.6)	58.1 (1.1)	64.5 (3.2)	69.3 (1.3)	60.9 (3.9)	63.6 (1.4)	63.7 (1.8)	67.9 (1.7)	54.9 (1.6)	57.7 (1.1)

data. After completing the training, the autoencoder is used to extract features from the testing data. Then using the extracted features of the training data we have trained all the machine learning algorithms available in the classification learner toolbox [21] in MATLAB. However, SVM and KNN with different kernels, and subspace discriminant of the ensemble method have better and consistent results. Therefore, we include the results of only the mentioned machine learning algorithms. The illustration of the framework is shown in Fig. 3. We repeat this process 10 times to ensure the results aren't biased by the data. Also, we carry out the experiments varying the threshold value T from 0 to 0.6 and using all the edges with positive PCC and all the edges with negative PCC. We use the accuracy (ACC) and AUC to evaluate the performance and standard deviation (stdv) to evaluate the stability. The results of the experiments are shown in Table 1.

From Table 1, we can see that higher classification results are achieved when a threshold is applied to the connectivity matrix. Both ACC and AUC are the highest for the threshold value of T = 0.4. However, a significant amount of information is lost when applying a large threshold, which is evident from the results of threshold T = 0.5 and T = 0.6. Also, the results are comparatively better when the information of both positive edges (edges with positive PCC) and negative edges (edges with negative PCC) are combined.

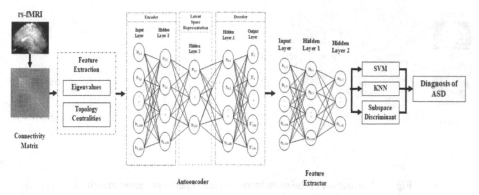

Fig. 3. Illustration of the main steps of using autoencoder as feature extractor

Table 2. Performance comparison of the DNN classifier with and without pre-training

Thresholding conditions	ACC% (stdv%)		AUC% (stdv%)	
	Without pre-training	With pre-training	Without pre-training	With pre-training
All edges with positive PCC	69.1 (7.4)	74.4 (1.4)	70.9 (10.3)	77.3 (1.3)
All edges with negative PCC	69.5 (0.6)	73.6 (1.1)	72.4 (1.0)	76.9 (0.9)
$T = 0$	69.4 (5.2)	73.8 (1.7)	70.8 (7.0)	75.2 (2.1)
$T = 0.1$	70.3 (3.4)	74.3 (1.3)	72.1 (4.1)	76.7 (0.9)
$T = 0.2$	**76.2 (4.1)**	**79.2 (0.8)**	**79.7 (0.7)**	**82.4 (0.8)**
$T = 0.3$	72.9 (1.0)	76.7 (1.1)	76.6 (1.4)	79.8 (1.2)
$T = 0.4$	74.6 (1.1)	77.1 (1.8)	77.7 (0.9)	80.9 (1.4)
$T = 0.5$	75.5 (1.2)	77.5 (1.1)	80.4 (1.4)	82.2 (0.9)
$T = 0.6$	75.8 (1.7)	77.3 (1.5)	80.7 (1.9)	81.7 (3.4)

Apart from using the autoencoder as a feature extractor, we also develop a neural network based classifier combined with an autoencoder. The architecture of our proposed neural network is shown in Fig. 2. In the neural network, the first two hidden layers are the same as the encoder and latent space representation of the autoencoder (Fig. 1). However, after the latent space representation instead of the decoder we have used two new layers (hidden layer 3′ and hidden layer 4′). Finally, the probability of data belonging to a particular class is calculated at the output layer.

To develop the DNN based classifier at first, we train the autoencoder in a similar way as the previous method. After completing the training of the autoencoder, we train the DNN (Fig. 2). The first two hidden layers of the neural network are pre-trained using the weights and biases of the first two hidden layers of the autoencoder. We use 10-fold cross-validation to train and evaluate the performance of the DNN classifier. In this

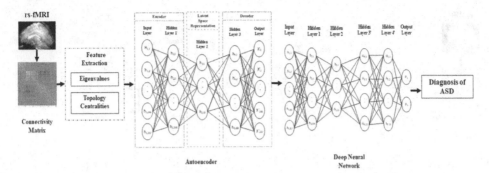

Fig. 4. Illustration of the main steps of neural network based classifier

Table 3. Performance analysis of features extracted from the combination of the pretrained autoencoder and the DNN

Thresholding conditions	Linear SVM (%) (stdv)		Medium Gaussian SVM (%) (stdv)		Coarse Gaussian SVM (%) (stdv)		Medium KNN (%) (stdv)		Cosine KNN (%) (stdv)		Weighted KNN (%) (stdv)		Subspace discriminant (%) (stdv)	
	ACC	AUC	ACC	AUC	ACC	AUC	ACC	AUC	ACC	AUC	ACC	AUC	ACC	AUC
All edges with positive PCC	62.8 (5.5)	66.1 (8.0)	66.3 (4.1)	70.1 (5.5)	55.3 (2.5)	60.8 (4.3)	67.8 (2.5)	71.4 (2.7)	66.1 (2.9)	69.9 (3.1)	67.7 (1.3)	72.4 (1.2)	60.7 (4.3)	63.3 (6.2)
All edges with negative PCC	66.0 (2.1)	71.4 (3.6)	66.6 (1.4)	74.2 (2.1)	63.7 (4.8)	68.1 (4.8)	66.8 (2.6)	72.4 (1.5)	66.7 (3.6)	72.3 (2.9)	67.2 (2.2)	73.8 (2.3)	66.6 (2.7)	70.9 (3.7)
T = 0	67.2 (4.4)	71.8 (4.8)	68.3 (2.1)	71.9 (2.2)	64.4 (6.7)	68.3 (7.0)	67.0 (2.6)	70.5 (3.1)	66.1 (3.4)	70.5 (2.5)	65.6 (2.9)	70.3 (3.0)	64.9 (4.1)	68.9 (5.3)
T = 0.1	67.0 (3.2)	71.2 (2.9)	69.1 (2.0)	74.7 (2.4)	66.6 (3.7)	71.2 (2.8)	68.4 (2.6)	74.5 (2.3)	67.9 (2.8)	76.0 (2.2)	67.9 (3.9)	75.1 (3.0)	67.1 (2.3)	71.1 (2.5)
T = 0.2	**72.8 (2.9)**	**76.3 (3.4)**	**73.7 (1.6)**	**76.7 (1.6)**	**73.0 (2.5)**	**77.9 (1.4)**	**73.2 (1.7)**	**77.7 (2.1)**	**74.6 (1.9)**	**78.7 (2.1)**	**72.7 (2.2)**	**77.2 (2.0)**	**74.4 (1.1)**	**77.8 (1.5)**
T = 0.3	66.1 (3.3)	71.7 (3.3)	66.4 (2.9)	71.9 (2.0)	63.0 (6.1)	67.7 (5.5)	67.1 (2.6)	72.8 (2.6)	66.9 (3.6)	72.1 (2.8)	66.4 (3.3)	72.9 (3.4)	67.8 (1.7)	71.6 (3.8)
T = 0.4	71.1 (4.7)	75.9 (3.8)	73.4 (1.1)	78.8 (1.0)	67.3 (3.1)	73.4 (3.1)	73.2 (2.5)	78.2 (1.8)	73.5 (1.2)	78.8 (1.1)	73.6 (1.5)	78.5 (1.5)	69.8 (3.8)	75.6 (2.8)
T = 0.5	64.4 (7.7)	69.9 (7.0)	70.3 (1.3)	73.1 (1.7)	60.2 (6.9)	64.9 (6.3)	67.7 (3.1)	72.5 (3.2)	67.2 (3.7)	72.8 (3.6)	70.6 (3.6)	76.7 (3.0)	64.0 (7.5)	68.5 (7.8)
T = 0.6	63.8 (3.1)	64.4 (3.2)	65.2 (3.8)	71.2 (3.0)	61.6 (2.8)	59.0 (1.9)	63.7 (1.8)	70.5 (1.5)	64.7 (1.9)	70.0 (2.2)	64.9 (2.6)	71.1 (2.5)	60.4 (3.0)	60.1 (1.9)

process of training, the DNN classifier is trained on 784 subjects, and the performance of the model is evaluated on the testing set of 87 subjects. The classification accuracy of the testing set is considered to be the accuracy of a particular fold. Finally, the average accuracy over all the folds is the accuracy of the model. We have repeated 10 times the process of training the autoencoder and 10-fold cross-validation of the DNN. Each time the subjects are selected randomly. The process is illustrated in Fig. 4.

To illustrate the effect of pre-training we also develop a DNN classifier without the pre-training. Apart from pre-training, everything else is kept the same. We repeat the 10-fold cross-validation ten times for the classifier. The comparison of the performance of the DNN with and without the pre-training is shown in Table 2.

From Table 2, the DNN without pre-training can achieve a classification accuracy of only 76.2%. However, the accuracy increases to 79.2% when pre-training is applied to the DNN. There is an increase in the classification accuracy for every thresholding condition. From the standard deviation in Table 2, it can be seen that the DNN with pre-training is more stable. However, in both cases, there is an increase of accuracy and AUC after applying some threshold, rather than using all the edges in the brain network and the results are maximum for the threshold of T = 0.2.

Table 4. Comparison of accuracy of proposed method and state of the art classification methods

Methods	Accuracy (%)
Heinsfeld et al. [4]	70.0
Eslami et al. [6]	70.1
Wong et al. [7]	71.1
Mostafa et al. [12]	77.7
Xing et al. [13]	66.8
Proposed autoencoder based feature extractor	74.6
Proposed autoencoder based DNN classifier	79.2

We use the data from both ASD and TC subjects when training the autoencoder. Even though the latent space representation creates a discriminate and salient representation of the input data, the difference in the features between different classes is not satisfactory (Table 1). However, when the DNN with pre-training is trained, the weights and biases of the hidden layers are updated to classify between ASD and TC subjects. As a result, the weights and biases of the encoder and latent space representation are also updated to create a more discriminate representation of the data. After completing the training of the DNN, the first two hidden layers can be used to extract a more discriminate representation of features. To enhance the performance of the feature extraction at first, we train the autoencoder. Then, we train the DNN with pre-training using the autoencoder. After completing the training of the DNN, we use the first two hidden layers to extract features and train different machine learning algorithms. Table 3 shows the performance of machine learning algorithms. Comparing Table 1 and Table 3 we can see that there is an increase in performance due to pre-training. The highest accuracy of 74.6% is achieved using the KNN classifier with the cosine kernel and the threshold value of T = 0.2.

Table 4 shows a comparison of the proposed study with other state-of-the-art methods. We have compared our study with only those studies, where the entire ABIDE 1 dataset is used. As we can see from Table 4, our proposed autoencoder based classifiers outperform all the previous studies.

4 Conclusion

In this paper, we have studied autoencoder based ASD diagnosis methods using the set of features proposed in [12]. In particular, we have pre-trained a neural network with an autoencoder. We have shown that the classification accuracy of the neural network has increased due to the pre-training. We have achieved a classification accuracy of 79.2% using the proposed diagnosis method, which is better than the state-of-the-art methods using the ABIDE 1 dataset. We have also proposed an autoencoder based feature selection method for the diagnosis of ASD. In this method, we have demonstrated that the learning of the DNN classifier can be incorporated in the autoencoder for dimensionality reduction. We have used traditional machine learning classifiers to evaluate the performance of the autoencoder based feature selection method and achieved a classification accuracy of 74.6%. In summary, our proposed diagnosis methods can diagnose ASD more accurately and precisely than state-of-the-art methods.

Acknowledgment. This work is supported by the Natural Science and Engineering Research Council of Canada (NSERC).

References

1. Hirvikoski, T., et al.: Premature mortality in autism spectrum disorder. Br. J. Psychiatry **208**(3), 232–238 (2016)
2. Lord, C., et al.: Autism diagnostic observation schedule: a standardized observation of communicative and social behavior. J. Autism Dev. Disord. **19**(2), 185–212 (1989). https://doi.org/10.1007/BF02211841
3. Lord, C., Rutter, M., Le Couteur, A.: Autism Diagnostic Interview-Revised: a revised version of a diagnostic interview for caregivers of individuals with possible pervasive developmental disorders. J. Autism Dev. Disord. **24**(5), 659–685 (1994). https://doi.org/10.1007/BF02172145
4. Heinsfeld, A.S., et al.: Identification of autism spectrum disorder using deep learning and the ABIDE dataset. NeuroImage Clin. **17**, 16–23 (2017)
5. Autism Brain Imaging Data Exchange I ABIDE I. http://fcon_1000.projects.nitrc.org/indi/abide/abide_I.html. Accessed 24 May 2019
6. Eslami, T., et al.: ASD-DiagNet: a hybrid learning approach for detection of Autism Spectrum Disorder using fMRI data. arXiv preprint arXiv:1904.07577v1
7. Wong, E., Anderson, J.S., Zielinski, B.A., Fletcher, P.T.: Riemannian regression and classification models of brain networks applied to autism. In: Wu, G., Rekik, I., Schirmer, M.D., Chung, A.W., Munsell, B. (eds.) CNI 2018. LNCS, vol. 11083, pp. 78–87. Springer, Cham (2018). https://doi.org/10.1007/978-3-030-00755-3_9
8. Kong, Y., et al.: Classification of autism spectrum disorder by combining brain connectivity and deep neural network classifier. Neurocomputing **324**, 63–68 (2019)
9. Watanabe, T., Rees, G.: Brain network dynamics in high-functioning individuals with autism. Nat. Commun. **8**(1), 16048 (2017)
10. Yahata, N., et al.: A small number of abnormal brain connections predicts adult autism spectrum disorder. Nat. Commun. **7**(1), 11254 (2016)
11. Arbabshirani, M.R., et al.: Single subject prediction of brain disorders in neuroimaging: promises and pitfalls. Neuroimage **145**, 137–165 (2017)

12. Mostafa, S., et al.: Diagnosis of autism spectrum disorder based on eigenvalues of brain networks. IEEE Access **7**(1), 128474–128486 (2019)
13. Xing, X., et al.: Convolutional neural network with element-wise filters to extract hierarchical topological features for brain networks. In: 2018 IEEE International Conference on Bioinformatics and Biomedicine, BIBM 2018, Madrid, pp. 780–783. IEEE (2019)
14. Martino, A.D., et al.: The autism brain imaging data exchange: towards a large-scale evaluation of the intrinsic brain architecture in autism. Mol. Psychiatry **19**(6), 659–667 (2014)
15. Cox, R.W.: AFNI: software for analysis and visualization of functional magnetic resonance neuroimages. Comput. Biomed. Res. **29**(3), 162–173 (1996)
16. Jenkinson, M., et al.: FSL. Neuroimage **62**(2), 782–790 (2012)
17. Power, J.D., et al.: Functional network organization of the human brain. Neuron **72**(4), 665–678 (2011)
18. Mijalkov, M., et al.: BRAPH: a graph theory software for the analysis of brain connectivity. PLoS ONE **12**(8), 0178798 (2017)
19. Hosseini-Asl, E., et al.: Alzheimer's disease diagnostics by adaptation of 3D convolutional network. In: 2016 IEEE International Conference on Image Processing (ICIP), USA, pp. 126–130. IEEE (2016)
20. Han, K., et al.: Autoencoder feature selector. arXiv preprint arXiv:1710.08310v1
21. Train Classification Models in Classification Learner App - MATLAB & Simulink. https://www.mathworks.com/help/stats/train-classification-models-in-classification-learner-app.html. Accessed 24 July 2019

FastFeatGen: Faster Parallel Feature Extraction from Genome Sequences and Efficient Prediction of DNA N^6-Methyladenine Sites

Md. Khaledur Rahman$^{(\boxtimes)}$

Indiana University Bloomington, Bloomington, IN 47408, USA
morahma@iu.edu

Abstract. N^6-methyladenine is widely found in both prokaryotes and eukaryotes. It is responsible for many biological processes including prokaryotic defense system and human diseases. So, it is important to know its correct location in genome which may play a significant role in different biological functions. Few computational tools exist to serve this purpose but they are computationally expensive and still there is scope to improve accuracy. An informative feature extraction pipeline from genome sequences is the heart of these tools as well as for many other bioinformatics tools. But it becomes reasonably expensive for sequential approaches when the size of data is large. Hence, a scalable parallel approach is highly desirable. In this paper, we have developed a new tool, called `FastFeatGen`, emphasizing both developing a parallel feature extraction technique and improving accuracy using machine learning methods. We have implemented our feature extraction approach using shared memory parallelism which achieves around $10\times$ speed over the sequential one. Then we have employed an exploratory feature selection technique which helps to find more relevant features that can be fed to machine learning methods. We have employed Extra-Tree Classifier (ETC) in `FastFeatGen` and performed experiments on rice and mouse genomes. Our experimental results achieve accuracy of 85.57% and 96.64%, respectively, which are better or competitive to current state-of-the-art methods. Our shared memory based tool can also serve queries much faster than sequential technique. All source codes and datasets are available at https://github.com/khaled-rahman/FastFeatGen.

Keywords: Genome sequence · Shared memory · Parallel feature extraction · Prediction model

1 Introduction

N^6-methyladenine (6mA) is very common in prokaryotes whose primary functions lie in the host defence system [1]. It is an abundant modification in mRNA

© Springer Nature Switzerland AG 2020
I. Măndoiu et al. (Eds.): ICCABS 2019, LNBI 12029, pp. 52–64, 2020.
https://doi.org/10.1007/978-3-030-46165-2_5

which has also been found in many multicellular eukaryotes such as *Caenorhabditis elegans* [2] and *Drosophila melanogaster* [3], and hence it has been proposed as a new epigenetic marker in eukaryotes [1]. Some studies have revealed that 6mA can control the acuity of infection and replication of RNA viruses like HIV and Zika virus [4,5]. A recent study demonstrates that 6mA modification can be heavily present in human genome and depletion of 6mA may lead to tumorigenesis [6]. Identification of 6mA in the genome will be helpful to characterize many biological functions and drug discovery.

Several experimental approaches exist to identify 6mA in genome. As described in [1], an antibody against N^6-methyladenine can identify N^6-adenine methylation in eukaryotic mRNAs which can further be used to identify N^6-methyladenine in DNA [7]. This technique is ambiguous due to the fact that other adenine-based modifications can be recognized. Liquid chromatography coupled with tandem mass spectrometry gives another comprehensive approach to identify 6mA [8]. Some restriction enzymes are sensitive to DNA methylation to differentiate between methylated nucleotides and unmethylated nucleotides which can be used to identify 6mA [9]. Single Molecule Real Time (SMRT) sequencing can determine kinetics of nucleotides during synthesis [10]. It has been applied to map 6mA and 5mC at the same time in *Escherichia coli* [11]. Noticeably, it can not differentiate between 6mA and 1mA, though this technique is very expensive. There are other experimental methods in the literature which have been found effective, e.g., liquid chromatography coupled with tandem mass spectrometry [2], and capillary electrophoresis alongside laser-induced fluorescence (CE & LIF) [12].

Most of the existing experimental methods are time consuming and expensive as mentioned above. Since the distribution of 6mA sites in the genome is not random and can follow some patterns, computational methods may be efficient and cost-effective. There are few such methods (6mA-Pred [13] and iDNA6mA-PseKNC [14]) which help to identify 6mA sites using supervised machine learning approaches. But, these methods adopt a sequential approach to extract features from DNA sequences which often slow down the process. Recently, convolutional neural networks (CNN) model has been applied to this problem [15]. However, comparison is not fair or ambiguous as jackknife testing is performed in 6mA-Pred whereas Tahir et al. use 20% samples of the dataset. Hence, we exclude this method from comparison due to inconsistency. There is still a need for a robust and precise tool that can facilitate faster and efficient identification of 6mA sites.

Various tools exist that extract features from DNA/protein sequences for prediction purposes, e.g., some tools extract features from genomic sequences to predict on-target activity in CRISPR/Cas9 technology [16,17] whereas other tools extract features from protein sequences to make efficient predictions [18–21]. But, almost all authors use a sequential approach [22–24] for feature extraction. With the advent of multi-core processors [25,26], a single machine can have two or more processing units which can lead to a significant speed-up of a properly written program. In this paper, we introduce such a parallelization technique in `FastFeatGen` (**Fast**er **Feat**ure extraction from **Gen**ome sequences) to extract features from DNA sequences which can also be applied to RNA/protein sequences as well with small modification. So, feature extraction techniques from large scale datasets will be significantly accelerated by our tool.

Over the years, a plethora of supervised machine learning based methods have been applied to solve several bioinformatics problems [27,28]. However, to the best of our knowledge, Chen et al. [13] were the first to tackle identification of 6mA sites in rice genome using Support Vector Machine (SVM). In this paper, we advance this concept with faster feature extraction and selection techniques and several supervised machine learning methods to achieve better performance. We also apply widely used neural network models to this problem in both supervised and unsupervised feature learning ways. Our extensive experimental results show that unsupervised way of CNN model is unable to surpass supervised one. This is likely due to the small size of the datasets, and, more interestingly, Extra-Tree Classifier (ETC) proposed by [29] performs very well despite its small set of features. We summarized our contributions as follows:

- We have introduced faster approaches for feature extraction techniques from genome sequences (Sect. 2.2).
- We have applied a lucid feature selection technique to select important and relevant features (Sect. 2.4).
- We have performed an extensive set of experiments using supervised machine learning methods to find a robust model (Sect. 2.3).
- We have compared our results with other state-of-the-art methods to show effectiveness and efficiency of our model (Sect. 3).
- We have also analyzed the processing time of new query sequences (Sect. 3.5).

2   Methods

The workflow of our tool is shown in Fig. 1. At first, features are extracted from input genome sequences (datasets). Then, relevant features are selected using the appropriate technique. After that, selected features are fed to machine learning methods to build the predictor. At this stage, several parameters are tuned until the model is optimized. Many existing methods follow Chou's 5-step rules (see [19]), and our tool is analogous to it. We describe each of these steps below.

2.1   Datasets

We have obtained two balanced datasets, Dataset1 and Dataset2 from [30] and [31], respectively. Dataset1 contains 880 positive samples (6mA sites) and 880 negative samples (non-6mA sites). Positive samples are from the rice genome, which are available at NCBI Gene Expression Omnibus[1] with the accession number GSE103145. Dataset2 contains 1934 positive samples and 1934 negative

---

[1] https://www.ncbi.nlm.nih.gov/geo/.

samples. Positive samples are curated from *Mus musculus* genome which are available in MethSMRT database [31]. Each sequence of both datasets is 41-bp long and nucleotide "A" is present at the center. More details about these datasets and negative samples generations can be found in [13] and [14].

We represent a 41-bp sequence by $S = s_1 s_2 s_3 \ldots s_{41}$, where s_i represents a nucleic acid in sequence S and $1 \leq i \leq 41$. Thus a dataset is represented by $D = S_1 S_2 S_3 \ldots S_{|D|}$, where $|D|$ is the size of dataset containing both positive and negative samples. For both datasets, $S_1 \ldots S_{|D|/2}$ are positive samples and $S_{|D|/2|+1} \ldots S_{|D|}$ are negative samples.

2.2   Feature Engineering

Features are the heart of supervised machine learning methods. We manually extract different types of information from given datasets and feed that information along with their corresponding true class to machine learning algorithms

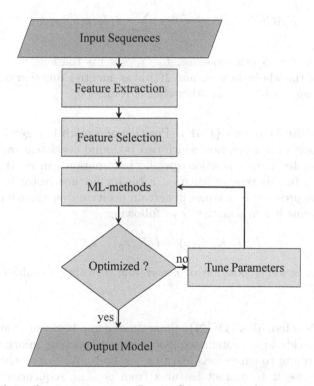

Fig. 1. Flowchart for `FastFeatGen`. Input Sequences - dataset containing DNA sequences, Feature Extraction - extract features from DNA sequences, Feature Selection - select relevant features using feature importance score, ML-methods - apply supervised machine learning methods on selected features, Optimized - check whether model is better or not, Tune Parameters - tune several parameters in ML-methods like learning rate, kernel function, etc., Output Model - produce optimized model for prediction.

for training and testing purposes. In this paper, we generate four types of features which are discussed below.

Nucleic Acids Composition (NAC). This is also known as position independent features or k-mers or n-grams. Each sequence may have certain short length patterns (also known as motifs) of NACs which are consistent over the whole dataset and so they may contribute to the learning model. In this technique, normalized frequency of a composition of nucleic acids is considered in corresponding sequence and finally a feature vector is constructed for the whole dataset. Length of a composition of nucleic acids is determined by order. For example, if order is 2, then all compositions of two nucleic acids is considered to extract features and a single feature vector is constructed for each composition. We normalize the frequency dividing by length of the sequence. We can define it mathematically as following:

$$NAC(K, S_j) = \frac{1}{L_{S_j} - k} \sum_{i=1}^{L_{S_j}-k} I(K, s_i \ldots s_{i+k})$$

where K is a k-mer, S is a sequence, L_{S_j} represents the length of S_j which is j^{th} sequence of the whole dataset, and $I(.)$ is an identity function which returns 1 when K is same as $s_i \ldots s_{i+k}$; otherwise, it returns 0.

Position Specific Features (PSF). Position of a motif in a genome sequence may carry important information which can be found consistent over the whole dataset. Undoubtedly, this position specific information can contribute to the learning model. In this type of features, a binary feature vector is constructed by checking the presence of a k-mer in certain position over the whole sequence. We can also define it mathematically as following:

$$PSF(K, S_j, p) = I(K, s_p \ldots s_{p+k})$$

where, p is the starting position of k-mer and all other variables carry same meaning as NAC.

Di-Gapped Nucleotides (DGN). Sometimes a gap between relative position of two amino acids in a protein sequence carry important information which may also contribute to supervised learning methods. We are motivated by this technique and use it to extract features from genome sequences as well. We construct a feature vector for each composition of two gapped nucleotides by normalizing its frequency in a sequence. We can formally define DGN as the following:

$$DGN(N_1, N_2, S_j, g) = \frac{1}{L_{S_j} - g} \sum_{i=1}^{L_{S_j}-g} I(s_i = N_1, s_{i+g} = N_2)$$

where, g is the gap length, N_1 and N_2 are two nucleotides. $I(.)$ is an identity function which returns 1 if i^{th} symbol of sequence S_j is N_1 and $(i+g)^{th}$ symbol of S_j is N_2; otherwise, it returns 0.

Bayesian Posterior Probability (BPP). In this technique, we first calculate the normalized frequency of each 2-mer for each position over the whole dataset. As we consider 2-mer, there are 40 different positions in a 41-bp sequence and we can have a total of 16 2-mers from all possible combinations of nucleotides. We construct a 40 × 16 matrix for the positive and negative samples separately. Then we extract BPP features in the following way: for each sequence we create a vector of size 80 where first 40 entries represent the posterior probabilities of position specific 2-mer in positive samples and last 40 entries represent the posterior probabilities in negative samples. More details of this approach can be found in [32].

Parallelization in Feature Extraction. We parallelize all the above feature extraction techniques in shared memory parallelism, which is accomplished through Single Instruction Multiple Data (SIMD) computing combined with multithreading. In this approach, instead of one sequence at a time, we pass nt sequences at a time to extract features using nt cores. Figure 2 shows a schematic diagram of our approach. A sequential feature extraction algorithm process one sequence at a time whereas `FastFeatGen` can process c sequences at a time using c available cores in a computing machine. It basically distributes all genome sequences to available cores and keeps it processing in parallel which results in faster running time.

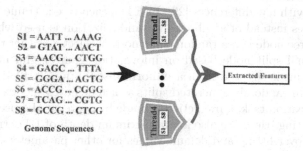

S1 = AATT ... AAAG
S2 = GTAT ... AACT
S3 = AACG ... CTGG
S4 = GAGC ... TTTA
S5 = GGGA ... AGTG
S6 = ACCG ... CGGG
S7 = TCAG ... CGTG
S8 = GCCG ... CTCG

Genome Sequences

Thread1 S1 ... S8
Thread4 S1 ... S8

Extracted Features

Fig. 2. A schematic diagram of parallel feature extraction where each thread constructs a specific feature using all input sequences.

2.3 Machine Learning Algorithms

In this tool, we use several supervised machine learning algorithms either to select informative features or to train the model. For all these approaches, we incorporate the popular **sklearn** python package in our tool unless otherwise mentioned explicitly. We provide short description for each of these models below.

Support Vector Machine (SVM). In computational genomics and proteomics, SVM is widely used for classification purpose. At first, the input dataset is transformed to high-dimensional feature space and then a kernel function maps the feature space to another dimension so that a boundary (also called margin) can separate the positive/negative classes. It creates a hyper-plane between positive and negative datasets so that margin between nearest positive samples and nearest negative samples is maximized. Nearest samples are often called support vectors and we can precisely state that the larger distances from hyper-plane equates to greater confidence levels in the predicted values. For SVM, we applied linear kernel function for feature selection and radial basis kernel function for classification purposes.

Random Forest (RF). RF is a popular ensemble method that is widely used for feature selection as well as for classification. Decision tree is the building block of RF which constructs rule sets over the feature space of training dataset, put class labels in the leaves of the tree, and branches denote conjunction of different rules that result in a corresponding class label. RF generally consists of a strategy to average a number of decision trees on various subsets of the dataset at training time to reduce variance and over-fitting. We allow maximum depth of RF trees to 500 in our model and use default values for other parameters.

Extra Tree Classifier (ETC). ETC is another ensemble method which is similar to RF with few differences [29]. In ETC, each tree is trained using whole training samples instead of a subset and randomization is used while top-down splitting of a tree node i.e., a random split node is selected rather than selecting a locally optimal split node based on information gain or gini impurity. This random cut-point is selected from a uniform distribution. A final split node is selected from all randomly generated splits which achieves the maximum score. For the classification task, a prediction is made by aggregated scores of each tree by majority voting. Here, we also allow maximum depth of ETC trees to 500 to avoid possible over-fitting and default values for other parameters.

Neural Networks (NN). NN is one of the most popular feed-forward methods being applied in different research fields including image processing, speech recognition, bioinformatics, etc. It consists of several cascaded layers (outputs of one layer are inputs to next layer) and each layer has a finite set of nodes (neurons in human brain). Each node of a layer is connected to all nodes to its next layer which results in a fully connected network. Each connection (edge) in the network represents a weight whose optimal value is learned by an iterative optimization algorithm like stochastic gradient descent. Each node in the network adds up the products of inputs and weights and passes through the activation function, which determines how much information should proceed further to influence the predictions.

We use a variation of the NN model called deep convolutional neural networks (CNN) which is very popular in computer vision. Unlike many conventional supervised learning processes, CNN does not require manually extracted features. Rather, it can extract features by itself, which is a large advantage for automating the classification process. Manually extracted features can also be fed to CNN to build a more diversified model. For CNN, we use one-hot encoding approach to represent a sequence where 'A', 'C', 'G' and 'T' are encoded as [1, 0, 0, 0], [0, 1, 0, 0], [0, 0, 1, 0] and [0, 0, 0, 1], respectively. As a result, each 41-bp sequence is represented by a 41×4 matrix.

2.4 Feature Selection

All extracted features do not contribute equally to build a better prediction model; in fact, some features do not contribute at all. We must find such irrelevant features and discard them from the feature list. We use SVM and RF for selecting and creating an important list of features that can help to train and optimize the prediction model. We use linear kernel of SVM and use a cutoff (threshold) value of 0.001 for each feature to be considered in our important feature list. Similarly, we select important features using RF based on its importance score. In the literature, RF is suggested to select a less biased or completely unbiased model [33], and many papers exist which use RF for feature selection. In our experiment, we discard any feature with zero importance from the important feature set.

2.5 Performance Evaluation

In the literature, cross-validation is a widely used technique to build a model which reduces selection bias and overfitting problems [33]. We perform 10-fold cross-validation and jackknife testing (also known as leave-one-out cross-validation) while performing experiments for training and testing our model. In 10-fold cross-validation, the dataset is partitioned into 10 equal folds. Among these, 9 are used to train the model whereas the remaining fold is used for testing purposes. This process is repeated 10 times with a different fold selected for testing each time. In jackknife testing, $n-1$ samples are used to train the model and the remaining sample is used for testing where n is the number of total samples in the dataset. This process is repeated n times so that each sample is considered once for testing. We use some notations of confusion matrix to define performance metrics in the following: we denote the total number of positive and negative samples by P and N, respectively; TP, TN, FP and FN represent the number of samples predicted as true positive, true negative, false positive, and false negative, respectively. We use four performance metrics, namely, Accuracy, Sensitivity, Specificity and Matthew's Correlation Coefficient which are denoted by Acc, Sn, Sp and MCC, respectively [20]. We express these performance metrics as following to compare our results with other tools.

$$Acc = \frac{TP + TN}{P + N}, Sn = \frac{TP}{P}, Sp = \frac{TN}{N}$$

$$MCC = \frac{TP * TN - FP * FN}{\sqrt{(TP + FP)(TP + FN)(TN + FP)(TN + FN)}}$$

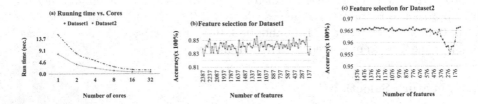

Fig. 3. (a) Running time vs. number of cores. (b) Accuracy for different sets of features in Dataset1. (c) Accuracy for different sets of features in Dataset2.

3   Results and Discussion

3.1   Experimental Setup

We perform all of our experiments in Haswell Compute Node of Cori super computer located in Berkeley lab which is configured as follows: each node has two sockets and each socket is populated with 16-core Intel®Xeon™Processor E5-2698 v3 ("Haswell") at 2.3 GHz, 32 cores per node, 36.8 Gflops/core and 128 GB DDR4 2.13 GHz memory. We wrote the coding for our feature extraction technique in C++ and the machine learning models in Python. Our tool requires at minimum GCC version 4.9, OpenMP version 4.5, and Python version 3.6.6. We provide source code with proper documentation, results and other information in our GitHub repository.

3.2   Parallel Feature Extraction Analysis

We use shared memory parallelism for feature extraction from genome sequences which is highly scalable. To extract features from both datasets, we set k, p and g for NAC, PSF and DGN as 5, 1...30, and 1...28, respectively. We show the results of run time for both datasets in Fig. 3(a). For different numbers of cores, running times are reported in seconds. We observe that when we increase the number of cores, running time decreases significantly and our tool extracts features using 32 cores for the above settings within a fraction of a second. `FastFeatGen` achieves a speed-up of around 10.3x for both Dataset1 and Dataset2 using 32 threads. Our tool can be applied to a wide range of biological sequence analysis problems for feature extraction where each given dataset contains a large number of DNA/RNA/Protein sequences.

3.3   Feature Importance Analysis

As discussed in Sect. 2.4, we select important features using linear SVM and RF. We use RF-based relevant feature selection for ETC and linear SVM-based feature selection for the SVM model. We show feature sets with accuracy for Dataset1 and Dataset2 using ETC model in Fig. 3(b) and Fig. 3(c), respectively. For Dataset1, our top-performing model contains 1237 features, among which 320 are from BPP; the rest of the features are mostly position specific. For Dataset2, our top-performing model contains 1326 features, which is higher than the case of Dataset1. 225 features are from BPP, and the rest of the features are mostly from PSF. For Dataset1, we see that C_26, A_27, and TA_26 are some of the features with higher importance scores. It indicates that Thymine in position 26 and Adenine in position 27 carry significant information for N^6-methyladenosine sites in the rice genome. On the other hand, G_21 and G_22 are two important features for Dataset2, which indicates that Guanine in positions 21 and 22 carry significant information for the mouse genome. We observe that the accuracy of the ETC model is always higher with its combined list of features rather than its individual features. So, we use combined list of features for prediction purposes.

3.4   Performance Analysis

Table 1. Comparison among different machine learning models for Dataset1.

Models	Accuracy	Models	Accuracy
SVM	93.06	NN	80.96
ETC	84.88	CNN	48.41

Comparison Among Different Learning Models. To build one efficient model from different machine learning algorithms discussed in Sect. 2.3, we conduct an extensive set of experiments. We compare SVM, ETC, NN and CNN using 10-fold cross-validation and observe that SVM and ETC models are competitive (see Table 1). SVM performs better than others because it uses a wide range of features, but its running time is very slow. On the other hand, ETC performs better than all other methods for a small set of features which also executes query much faster than others. The performance of NN method is comparative to SVM and ETC while CNN is the worst performer. As both datasets are small in size, CNN can not utilize its automatic feature extractions approach in depth and hence shows worse performance. We select ETC model as a representative of FastFeatGen for comparison with other state-of-the-art tools.

Comparison with Existing Tools. We compare our model with 6mA-Pred and PseDNC, which are considered as the state-of-the-art methods for Dataset1. Following the trend of 6mA-Pred, we generate all results using jackknife test. From Table 2, we see that `FastFeatGen` (with 1237 features) achieves an accuracy of 85.56%, which is better than other existing tools. It also achieves higher Specificity and MCC which are better than other tools. Furthermore, `FastFeatGen` still outperforms existing tools in terms of Accuracy, Specificity and MCC with only 187 features.

Table 2. Comparison among different tools for Dataset1.

Tools\Metrics	Accuracy	Sensitivity	Specificity	MCC
`FastFeatGen` (1237 features)	**85.56**	81.47	**89.65**	**0.71**
`FastFeatGen` (187 features)	85.45	81.81	89.09	0.71
6mA-Pred	83.13	**82.95**	83.3	0.66
PseDNC	64.55	63.52	65.57	0.29

For Dataset2, we compare our method with iDNA6mA-PseKNC, which is the only tool for the mouse genome. Here, we also perform jackknife testing following the trend of iDNA6mA-PseKNC. From Table 3, we see that `FastFeatGen` is better or very competitive in all metrics.

Table 3. Comparison among different tools for Dataset2.

Tools\Metrics	Accuracy	Sensitivity	Specificity	MCC
`FastFeatGen` (101 features)	96.63	**93.49**	100	**0.94**
iDNA6mA-PseKNC	**96.73**	93.28	100	0.93

3.5 Query Time Analysis

The general purpose of building a machine learning model is to make prediction for more unknown genome sequences, which is expected to be faster. Most of the sequence analysis tools or web-servers can not provide such facility, or authors impose restrictions on the number of query sequences. `FastFeatGen` provides a scalable solution to this problem which has no restrictions. Users can query as many sequences as they want. We again employ parallel feature extraction technique here and enable parallel job processing of ETC in **sklearn** package. In summary, `FastFeatGen` can serve 200 queries within 0.7 second(s) using 32 threads.

4 Conclusions

In this paper, we have introduced a novel tool called `FastFeatGen` which uses multi-core processing for faster extraction of features from genome sequences. Then, it performs lucid feature selection techniques that select high quality features to feed into machine learning methods. Finally, we build a precise model using extra tree classifier which performs very well using a small set of features. We have shown that our tool performs better than state-of-the-art methods on two publicly available datasets. `FastFeatGen` achieves an accuracy of 85.56% and 96.63% for rice and mouse genomes, respectively, which are superior or competitive to current state-of-the-art methods. Our tool can predict for a wide range of new query sequences within fraction of a second which is clearly an advantage over web-server based tools.

Our future goal is to improve and apply our faster feature extraction techniques to other biological problems that involves protein or RNA sequences. We also would like to include more feature extraction techniques to `FastFeatGen`.

References

1. Luo, G.-Z., Blanco, M.A., Greer, E.L., He, C., Shi, Y.: DNA N^6-methyladenine: a new epigenetic mark in eukaryotes? Nat. Rev. Mol. Cell Biol. **16**(12), 705 (2015)
2. Greer, E.L., et al.: DNA methylation on N^6-adenine in *C. elegans*. Cell **161**(4), 868–878 (2015)
3. Zhang, G., et al.: N^6-methyladenine DNA modification in *Drosophila*. Cell **161**(4), 893–906 (2015)
4. Lichinchi, G., et al.: Dynamics of the human and viral m^6A RNA methylomes during HIV-1 infection of T cells. Nat. Microbiol. **1**(4), 16011 (2016)
5. Lichinchi, G., et al.: Dynamics of human and viral RNA methylation during Zika virus infection. Cell Host Microbe **20**(5), 666–673 (2016)
6. Xiao, C.-L., et al.: N^6-methyladenine DNA modification in the human genome. Mol. Cell **71**(2), 306–318 (2018)
7. Fu, Y., et al.: N^6-methyldeoxyadenosine marks active transcription start sites in *Chlamydomonas*. Cell **161**(4), 879–892 (2015)
8. Frelon, S., Douki, T., Ravanat, J.-L., Pouget, J.-P., Tornabene, C., Cadet, J.: High-performance liquid chromatography- tandem mass spectrometry measurement of radiation-induced base damage to isolated and cellular DNA. Chem. Res. Toxicol. **13**(10), 1002–1010 (2000)
9. Roberts, R.J., Macelis, D.: Rebase—restriction enzymes and methylases. Nucleic Acids Res. **29**(1), 268–269 (2001)
10. Flusberg, B.A., et al.: Direct detection of DNA methylation during single-molecule, real-time sequencing. Nat. Methods **7**(6), 461 (2010)
11. Fang, G., et al.: Genome-wide mapping of methylated adenine residues in pathogenic *Escherichia coli* using single-molecule real-time sequencing. Nat. Biotechnol. **30**(12), 1232 (2012)
12. Krais, A.M., Cornelius, M.G., Schmeiser, H.H.: Genomic N^6-methyladenine determination by MEKC with LIF. Electrophoresis **31**(21), 3548–3551 (2010)
13. Chen, W., Lv, H., Nie, F., Lin, H.: i6mA-Pred: identifying DNA N^6-methyladenine sites in the rice genome. Bioinformatics **35**(16), 2796–2800 (2019)

14. Feng, P., Yang, H., Ding, H., Lin, H., Chen, W., Chou, K.-C.: iDNA6mA-PseKNC: identifying DNA N^6-methyladenosine sites by incorporating nucleotide physico-chemical properties into PseKNC. Genomics **111**(1), 96–102 (2019)

15. Tahir, M., Tayara, H., Chong, K.T.: iDNA6mA (5-step rule): identification of DNA N^6-methyladenine sites in the rice genome by intelligent computational model via Chou's 5-step rule. Chemometrics and Intelligent Laboratory Systems (2019)

16. Doench, J.G., et al.: Optimized sgRNA design to maximize activity and minimize off-target effects of CRISPR-Cas9. Nat. Biotechnol. **34**(2), 184 (2016)

17. Rahman, M.K., Rahman, M.S.: CRISPRpred: a flexible and efficient tool for sgR-NAs on-target activity prediction in CRISPR/Cas9 systems. PLoS ONE **12**(8), e0181943 (2017)

18. Manavalan, B., Lee, J.: SVMQA: support–vector-machine-based protein single-model quality assessment. Bioinformatics **33**(16), 2496–2503 (2017)

19. Chou, K.-C.: Some remarks on protein attribute prediction and pseudo amino acid composition. J. Theor. Biol. **273**(1), 236–247 (2011)

20. Rahman, M.S., Rahman, M.K., Kaykobad, M., Rahman, M.S.: isGPT: an optimized model to identify sub-Golgi protein types using SVM and Random Forest based feature selection. Artif. Intell. Med. **84**, 90–100 (2018)

21. Rahman, M.S., Rahman, M.K., Saha, S., Kaykobad, M., Rahman, M.S.: Antigenic: an improved prediction model of protective antigens. Artif. Intell. Med. **94**, 28–41 (2019)

22. Cao, D.-S., Xu, Q.-S., Liang, Y.-Z.: propy: a tool to generate various modes of Chou's PseAAC. Bioinformatics **29**(7), 960–962 (2013)

23. Liu, B., Liu, F., Fang, L., Wang, X., Chou, K.-C.: repDNA: a Python package to generate various modes of feature vectors for DNA sequences by incorporating user-defined physicochemical properties and sequence-order effects. Bioinformatics **31**(8), 1307–1309 (2014)

24. Liu, B.: BioSeq-Analysis: a platform for DNA, RNA and protein sequence analysis based on machine learning approaches. Brief. Bioinform. (2017)

25. Schauer, B.: Multicore processors–a necessity. In: ProQuest Discovery Guides, pp. 1–14 (2008)

26. Blake, G., Dreslinski, R.G., Mudge, T.: A survey of multicore processors. IEEE Signal Process. Mag. **26**(6), 26–37 (2009)

27. Larranaga, P., et al.: Machine learning in bioinformatics. Brief. Bioinform. **7**(1), 86–112 (2006)

28. Stephenson, N., et al.: Survey of machine learning techniques in drug discovery. Curr. Drug Metab. **20**(3), 185–193 (2019)

29. Geurts, P., Ernst, D., Wehenkel, L.: Extremely randomized trees. Mach. Learn. **63**(1), 3–42 (2006). https://doi.org/10.1007/s10994-006-6226-1

30. Zhou, C., et al.: Identification and analysis of adenine N^6-methylation sites in the rice genome. Nat. Plants **4**(8), 554 (2018)

31. Ye, P., Luan, Y., Chen, K., Liu, Y., Xiao, C., Xie, Z.: MethSMRT: an integrative database for DNA N6-methyladenine and N4-methylcytosine generated by single-molecular real-time sequencing. Nucleic Acids Res. **45**, D85–D89 (2016). https://doi.org/10.1093/nar/gkw95

32. Shao, J., Xu, D., Tsai, S.-N., Wang, Y., Ngai, S.-M.: Computational identification of protein methylation sites through bi-profile Bayes feature extraction. PLoS ONE **4**(3), e4920 (2009)

33. Cawley, G.C., Talbot, N.L.: On over-fitting in model selection and subsequent selection bias in performance evaluation. J. Mach. Learn. Res. **11**, 2079–2107 (2010)

Optimized Multiple Fluorescence Based Detection in Single Molecule Synthesis Process Under High Noise Level Environment

Hsin-Hao Chen$^{(\boxtimes)}$ (iD) and Chung-Chin Lu

Department of Electrical Engineering,
National Tsing Hua University, Hsinchu 30013, Taiwan
andrechen0513@gmail.com, cclu@ee.nthu.edu.tw

Abstract. Single molecule sequencing contributes to overall human advancement in the areas including but not limited to genomics, transcriptomics, clinical test, drug development, and cancer screening. Furthermore, fluorescence based sequencing is mostly employed in single molecule sequencing among other methods, specifically in the fields of DNA sequencing. Contemporary fluorescence labeling methods utilize a Charge-coupled Device camera to capture snapshots of multiple pixels on the single molecule sequencing. We propose a method for fluorescence labeling detection with a single pixel, which excels in high accuracy and low resource requirement in the low signal-to-noise ratio conditions. Such a method also benefits from higher throughput compared to others. The context in this study explores the single molecule synthesis process modeling using negative binomial distributions. Also, including the method of maximum likelihood and Viterbi algorithm in this modeling improves signal detection accuracy. The fluorescence-based model is most beneficial to simulate actual experiment processes and to facilitate in understanding the relations between fluorescence emission and signal receiving event.

Keywords: Genomic sequencing · Fluorescence labeling · Fluorescence based model · Signal detection · Single molecule synthesis process · Maximum likelihood · Viterbi algorithm

1 Introduction

As technology advances at an exponential rate, the distinctive field of biological science has been in expansion specifically in nucleic acid detection, flourishing into various analytical methods and strategies [11]. Evidently, the performance of genomic sequencing detection also rises with new available nucleic acid detection methods. Fluorescence signaling method has been one of the most efficient tools in its league to explore within vast genomes [9]. Advantages include but not

© Springer Nature Switzerland AG 2020
I. Măndoiu et al. (Eds.): ICCABS 2019, LNBI 12029, pp. 65–76, 2020.
https://doi.org/10.1007/978-3-030-46165-2_6

limited to an increased level of sensitivity, multiplexing capabilities, simultaneous detection on fluorescence properties [3].

Numerous agent supplementing methods are employed to identify organic molecules such as proteins and nucleic acids and to augment sensitive detectability in the controlled assay. Fluorescence labeling excels in its ranks in which it provides detectable sensitivity with light signal exposure when stimulated during sequencing [8]. Additionally, each fluorescence agent has its own unique fluorescence property or light wavelength signature, thus the detection devices are able to conduct the simultaneous observations on more than one molecule, saving time resources and cutting down assay duration [10].

A molecule under a fluorescent event is excited and emits light at a different wavelength than it was exposed to. Through the photoelectric effect, detection devices, such as camera equipped with Charge-coupled Device (CCD) capability or a group of photodiodes, are able to monitor an increased electric voltage or bigger RGB values as fluorescent light emitted from the compound [13]. As from the conventional fluorescence detection methods, signal-to-noise ratio (SNR) has always been a bottleneck to data collection and integrity [7]. In such schemes, the algorithm embedded in a CCD camera relies on optimized extraction and signal strength comparison to increase SNR [6]. A matrix of pixels is mandatory to accomplish such results. Even so, higher error rate, higher cost per base, and lower throughput associated with such method are at a disadvantage comparing to other techniques [1]. The consensus sequencing method is employed to compensate for the high error rate [5]. However, it requires higher computing storage, more complicated template, and more time resulting from repetitive iterations.

In the scenario described in this study, polymerase synthesis and fluorescent emission are indispensable elements in single molecular sequencing. To improve the signal integrity analysis and cost per base in sequencing, statistical modeling is introduced to simulate polymerase synthesis and the fluorescent emission process. Within such modeling, an algorithm containing the method of maximum likelihood and Viterbi algorithm [14] measures and distinguishes the distinct fluorescences in low SNR conditions, together with a single, three-junction photodiode which records the RGB values of light emission [15].

2 Materials and Models

2.1 The Single Molecule Synthesis Process

The DNA polymerase synthesizes the DNA nucleotides in a sequence similar to the single molecule synthesis process (SMSP) as in the previous study [2]. SMSP is modeled as a discrete-time stochastic process $\{Z(t), t \geq 1\}$ on a state space $\{A, T, G, C, d\}$, where state A (T, G or C) indicates that a dATP (dTTP, dGTP or dCTP) is being incorporated by the DNA polymerase and state d indicates that no dNTP is being incorporated.

Let S_n and $T_n - 1$ be the start time and the stop time of the nth incorporation of dNTP for $n \geq 1$. Then

$$0 = T_0 \leq S_1 < T_1 \leq S_2 < T_2 \leq \cdots \leq S_n < T_n \leq \cdots$$

Let $X_n \equiv Z(S_n)$ be the nth nucleotide to be incorporated. The length of the nth interpulse duration is $W_n = S_n - T_{n-1}$ and the length of the nth incorporation period or pulse width (herebelow indicated as pulse width) is $Y_n = T_n - S_n$. Due to the lack of structural information of a short piece of DNA, it can be assumed that $\{X_n, n \geq 1\}$ is an independent and identically distributed (IID) sequence of random variables with uniform distribution over $\{A, T, G, C\}$, and we will model the distributions of W_n and Y_n by negative binomial distributions with parameters (l, q) and (r, p) respectively:

$$P(W_n = k) = C_{l-1}^{k-1} q^l (1-q)^{k-l}, k = l, l+1, l+2, \ldots. \tag{1}$$

$$P(Y_n = k) = C_{r-1}^{k-1} p^r (1-p)^{k-r}, k = r, r+1, r+2, \ldots. \tag{2}$$

With the above assumptions, the alternating process $\{(W_n, Y_n), n \geq 1\}$ becomes a discrete-time Markov chain $\{L(t), t \geq 1\}$ with the state space

$$\mathcal{S} = \{d_1, d_2, \ldots, d_l, A_1, A_2, \ldots, A_r, T_1, T_2, \ldots, T_r, G_1, G_2, \ldots, G_r, C_1, C_2, \ldots, C_r\}$$

of size $4r + l$. Let $S_i = \{A_i, T_i, G_i, C_i\}$. The state transition diagram of the Markov chain $\{L(t), t \geq 1\}$ is in Fig. 1(a) and the state transition probabilities are

$$\pi_{d_i,s} = \begin{cases} 1-q, & \text{if } s = d_i, \\ q, & \text{if } s = d_{i+1}, \text{ for } 1 \leq i \leq l-1, \\ 0, & \text{otherwise}, \end{cases} \tag{3}$$

$$\pi_{d_l,s} = \begin{cases} 1-q, & \text{if } s = d_l, \\ \frac{q}{4}, & \text{if } s \in S_1, \\ 0, & \text{otherwise}, \end{cases} \tag{4}$$

$$\pi_{A_i,s} = \begin{cases} 1-p, & \text{if } s = A_i, \\ p, & \text{if } s = A_{i+1}, \text{ for } 1 \leq i \leq k-1, \\ 0, & \text{otherwise}, \end{cases} \tag{5}$$

$$\pi_{A_k,s} = \begin{cases} 1-p, & \text{if } s = A_k, \\ p, & \text{if } s = d_1, \\ 0, & \text{otherwise}. \end{cases} \tag{6}$$

With the assumption in Eq. (2), the transition probabilities, $\pi_{A_i,s}$, $\pi_{T_i,s}$, $\pi_{G_i,s}$, and $\pi_{C_i,s}$ are identical.

2.2   The Emission Process

When a dNTP is being incorporated by the DNA polymerase, a fluorescence light will emit and be detected by a three-junction photodiode. While no dNTP is being incorporated, only ambient light is detected, and of course, ambient light also exists in an incorporation period. Consequently, the photodetector will output the fluorescence plus ambient light intensity signal during an incorporation period and the ambient light intensity signal only outside of the incorporation period.

Let $\{E(t), t \geq 1\}$ be the output intensity signal of the photodetector associated with the SMSP $\{Z(t), t \geq 1\}$, called the emission process. Assume that the emission vector $E(t)$ depends only on the state $Z(t)$ at time t. Note that $\{Z(t), t \geq 1\}$ represents the SMSP $\{(W_n, Y_n, X_n), n \geq 1\}$ and

$$Z(t) = \begin{cases} d, & \text{if } \sum_{i=1}^{n-1}(W_i + Y_i) < t \leq \sum_{i=1}^{n-1}(W_i + Y_i) + W_n, \\ X_n, & \text{if } \sum_{i=1}^{n-1}(W_i + Y_i) + W_n < t \leq \sum_{i=1}^{n}(W_i + Y_i). \end{cases}$$

The Ambient Light Intensity Signal. Under the conditions that the ambient light source is originated from the dNTPs and the interpulse duration is in the steady state, the ambient light intensity signal can be modeled by a constant signal vector $a_{Z(t)}$, which is undetermined and will be estimated for the photodetector in a pixel.

$$a_x = \begin{cases} a_d, & \text{if } x = d, \\ a_A, & \text{if } x = A, \\ a_T, & \text{if } x = T, \\ a_G, & \text{if } x = G, \\ a_C, & \text{if } x = C. \end{cases}$$

The Fluorescence Light Intensity Signal. The fluorescence light intensity signal during an incorporation period depends on the dye molecule bound with the nucleotide under synthesis as well as the distance between the dye molecule and the photodiode. The biochemical reactions take place in a closed space where fluorescence light can be fully captured by the photodetector. Such a closed space is called a synthesis well.

Assume that the DNA polymerase has a fixed position in the synthesis well during an incorporation period so that the captured dNTP plus dye molecule by the enzyme has a constant distance from the photodiode. Thus if $Z(t) = x$ for $t \in [S_n, T_n - 1]$, where $x \in \{A, T, G, C\}$, then the fluorescence light intensity signal will be $\Upsilon_t s_x, t \geq 1$, where s_x is the detected signal vector of the three-junction photodiode to the fluorescence light emitted from the dye molecule bound with a dxTP in a nominal distance from the photodiode. The above assumption derives from our observations that the photoelectric effect is linear within the fluorescent reactions.

Υ_t is the fading coefficient resulting from the variation of the true distance from the dye molecule to the photodiode in the time t of the incorporation period relative to the nominal distance. Since limitations exist in the photodetector and the sequencing reactions, the emission intensity $\Upsilon(t)$ has the minimum and maximum values $(I_{A,\min}, I_{A,\max})$, $(I_{T,\min}, I_{T,\max})$, $(I_{G,\min}, I_{G,\max})$, $(I_{C,\min}, I_{C,\max})$ when synthesizing ATP, TTP, GTP, CTP respectively. Each characteristic signal vector s_x, $x \in \{A, T, G, C\}$, is an undetermined parameter vector and will be estimated for each of the photodetectors. $\{\Upsilon(t), t \geq 1\}$ is the fluorescence intensity process in photon/ms and can be modeled as a wide-sense stationary or cyclostationary process depending on the movement kinematics of the DNA polymerase in the synthesis well.

The Emission During an Interpulse Duration. The emission signal during the nth interpulse duration $[T_{n-1}, S_n - 1]$ is

$$E(t) = a_d, \ t \in [T_{n-1}, S_n - 1].$$

The Emission During an Incorporation Period. The emission signal during the nth incorporation period $[S_n, T_n - 1]$ is

$$E(t) = \Upsilon(t)s_{Z(t)} + a_{Z(t)}, \ t \in [S_n, T_n - 1].$$

2.3   The Received Process

The photodetector, as well as the readout circuit, will introduce noise to the emission process. From our noise measurement of the photodetector, the cross-correlations between three junctions are close to zero and therefore negligible to calculation. We model this noise as a white Gaussian vector process $\{N(t), t \geq 1\}$ with covariance matrix

$$\Lambda = \begin{bmatrix} \sigma_1^2 & 0 & 0 \\ 0 & \sigma_2^2 & 0 \\ 0 & 0 & \sigma_3^2 \end{bmatrix},$$

where σ_i^2 is the average noise power of the ith junction channel and will be estimated for each pixel. The joint probability density function (jpdf) of the noise vector $N(t)$ at time t is

$$f_N(n) = \frac{1}{\sqrt{(2\pi)^3 |\Lambda|}} e^{-\frac{1}{2} n^t \Lambda^{-1} n} = \prod_{i=1}^{3} \frac{1}{\sqrt{2\pi\sigma_i^2}} e^{-\frac{n_i^2}{2\sigma_i^2}},$$

where $|\Lambda|$ is the determinant of the covariance matrix Λ. The noise process $\{N(t), t \geq 1\}$ is assumed to be independent of the SMSP $\{Z(t), t \geq 1\}$ and the fading process $\{\Upsilon(t), t \geq 1\}$.

We assume that the noise process $\{N(t), t \geq 1\}$ is additive to the emission process $\{E(t), t \geq 1\}$ so that the received signal process $\{R(t), t \geq 1\}$ is

$$R(t) = E(t) + N(t), \ t \geq 1 \tag{7}$$

$$= \begin{cases} a_d + N(t), & \text{if } Z(t) = d, \\ \Upsilon(t)s_A + a_A + N(t), & \text{if } Z(t) = A, \\ \Upsilon(t)s_T + a_T + N(t), & \text{if } Z(t) = T, \\ \Upsilon(t)s_G + a_G + N(t), & \text{if } Z(t) = G, \\ \Upsilon(t)s_C + a_C + N(t), & \text{if } Z(t) = C. \end{cases} \tag{8}$$

3   Methods and Algorithms

In this section, the focus is positioned on the decoding and the parameter estimation algorithms. As a first phase, the initial pixel parameters are prepared by the default values or the values calculated from the experiment data. Next phase, the entire sequence is detected by the decoding algorithm and the default parameters.

1. **Initial Phase**: The initial pixel parameters s_A, s_T, s_G, s_C, a_A, a_T, a_G, a_C, a_d and σ are given.
2. **Decoding Phase**: $\{I(t), t \geq 1\}$ and $\{\alpha(t), t \geq 1\}$ of the alternating process $\{L(t), t \geq 1\}$ and the fluorescence intensity process $\{\Upsilon(t), t \geq 1\}$ are obtained by the known training nucleotide sequence $\{x_n, n \geq 1\}$ and the given estimated pixel parameters s_A, s_T, s_G, s_C, a_A, a_T, a_G, a_C, a_d and σ.

3.1   Estimation of the Initial Pixel Parameters

The initial pixel parameters s_A, s_T, s_G, s_C, a_A, a_T, a_G, a_C, a_d and σ are from the experimental measurement data. For the estimation of the parameters, a simple linear regression is used, $R(t) = \Upsilon(t)s_i + a_i + N(t), i \in \{d, A, T, G, C\}$ and $s_d = 0$, which is equivalent to Eq. (8). Moreover, the R-squared (R^2) is calculated to determine if the simple linear regression is well fitted. σ is calculated by the measurement in the interpulse duration.

3.2   Decoding Phase

Assume that the estimated pixel parameters s_A, s_T, s_G, s_C, a_A, a_T, a_G, a_C, a_d and σ are given. The likelihood function $f_{R|Z,\Upsilon}(r|z, \alpha)$ of the received signal $\{R(t), t \geq 1\}$ to be $\{r(t), t \geq 1\}$, given the SMSP $\{Z(t), t \geq 1\}$ to be $\{z(t), t \geq 1\}$ and the emission intensity process $\{\Upsilon(t), t \geq 1\}$ to be $\{\alpha(t), t \geq 1\}$, is

$$f_{R|Z,\Upsilon}(r|z, \alpha) = \prod_{t \geq 1} f_{R(t)|Z(t),\Upsilon(t)}(r(t)|z(t), \alpha(t))$$

since the noise process $\{N(t), t \geq 1\}$ is a white Gaussian process, where

$$f_{R(t)|Z(t),\Upsilon(t)}(r(t)|z(t), \alpha(t))$$
$$= \begin{cases} \prod_{j=1}^{3} \frac{1}{\sqrt{2\pi}\sigma_j} e^{-\frac{1}{2}(\tau_j(t))^2}, \text{if } z(t) = d, \\ \prod_{j=1}^{3} \frac{1}{\sqrt{2\pi}\sigma_j} e^{-\frac{1}{2}(\varphi_{x,j}(t))^2}, \text{if } z(t) = x \in \{A, T, G, C\} \end{cases}$$

where $\tau_j(t) = \frac{r_j(t) - a_{d,j}}{\sigma_j}$

and $\varphi_{x,j}(t) = \frac{r_j(t) - a_{x,j} - \alpha(t)s_{x,j}}{\sigma_j}$ for $z(t) = x \in \{A, T, G, C\}$.

Now given the estimated pixel parameters s_A, s_T, s_G, s_C, a_A, a_T, a_G, a_C, a_d and σ, the decoded versions $\{z(t), t \geq 1\}$ and $\{\alpha(t), t \geq 1\}$ of the SMSP

$\{Z(t), t \geq 1\}$ and the emission intensity process $\{\Upsilon(t), t \geq 1\}$ can be obtained by the method of maximum likelihood (ML),

$$(z, \alpha) = \arg \max_{(z,\alpha)} \sum_{t \geq 1} \ln f_{R(t)|Z(t),\Upsilon(t)} (r(t)|z(t), \alpha(t)),$$

where $\ln f_{R(t)|Z(t),\Upsilon(t)} (r(t)|z(t), \alpha(t))$

$$= \begin{cases} \sum_{j=1}^{3} \ln \frac{1}{\sqrt{2\pi}\sigma_j} - \frac{1}{2} \sum_{j=1}^{3} (\tau_j(t))^2, & \text{if } z(t) = d, \\ \sum_{j=1}^{3} \ln \frac{1}{\sqrt{2\pi}\sigma_j} - \frac{1}{2} \sum_{j=1}^{3} (\varphi_{x,j}(t))^2, & \text{if } z(t) = x \in U \end{cases}$$

and $U = \{A, T, G, C\}$. Since the term $\sum_{j=1}^{3} \ln \frac{1}{\sqrt{2\pi}\sigma_j}$ is irrelevant to the maximization process, we will define a metric $m(z(t), \alpha(t))$ as follows:

$$m(z(t), \alpha(t)) = \begin{cases} \sum_{j=1}^{3} (\tau_j(t))^2, & \text{if } z(t) = d, \\ \sum_{j=1}^{3} (\varphi_{x,j}(t))^2, & \text{if } z(t) = x \in U. \end{cases}$$

Then the ML decoded version of $\{Z(t), t \geq 1\}$ and $\{\Upsilon(t), t \geq 1\}$ is

$$(z, \alpha) = \arg \min_{z} \sum_{t \geq 1} \min_{\alpha(t)} m(z(t), \alpha(t)).$$

Subsequently, given a hypothetical SMSP $\{z(t), t \geq 1\}$, the minimization of the sum $\sum_{t \geq 1} m(z(t), \alpha(t))$ of metrics $m(z(t), \alpha(t))$ over the emission intensity process $\{\alpha(t), t \geq 1\}$ can be done by the minimization of the metrics $m(z(t), \alpha(t))$ over the intensity $\alpha(t)$ at each time t.

Let $\alpha^*(t|z(t)) = \arg \min_{\alpha(t)} m(z(t), \alpha(t))$, then

$$\alpha^*(t|z(t)) = \begin{cases} \text{unknown}, & \text{if } z(t) = d, \\ Q\left(I_{x,\min}, \alpha^\#(t|z(t)), I_{x,\max}\right), & \text{if } z(t) = x \in U, \end{cases}$$

where for $a < c$,

$$Q(a, b, c) = \begin{cases} a, & \text{if } b < a, \\ b, & \text{if } a \leq b \leq c, \\ c, & \text{if } b > c, \end{cases}$$

and

$$\alpha^\#(t|z(t)) = \begin{cases} \text{unknown}, & \text{if } z(t) = d, \\ \frac{\sum_{j=1}^{3} (\varphi_{x,j}(t))(s_{x,j}/\sigma_j)}{\sum_{j=1}^{3} (s_{x,j}/\sigma_j)^2}, & \text{if } z(t) = x \in U, \end{cases}$$

by doing the minimization of the metric $m(z(t), \alpha(t))$ over all $\alpha(t) \in \mathbb{R}$.

$$m^*(z(t)) = \begin{cases} \sum_{j=1}^{3} (\tau_j(t))^2, & \text{if } z(t) = d, \\ \sum_{j=1}^{3} \left((\varphi_{x,j}(t)) - \alpha^*(t|z(t)) \left(\frac{s_{x,j}}{\sigma_j}\right)\right)^2, & \text{if } z(t) = x \in U. \end{cases} \tag{9}$$

Now the ML decoded versions $\{z(t), t \geq 1\}$ and $\{\alpha(t), t \geq 1\}$ of the SMSP $\{Z(t), t \geq 1\}$ and the fluorescence intensity process $\{\Upsilon(t), t \geq 1\}$ are

$$(z, \alpha) = \arg \min_{(z, \alpha^{\#}(t|z(t)))} \sum_{t \geq 1} m^*(z(t)). \tag{10}$$

Next, the modified Viterbi algorithm is applied to solve this minimization problem in the above equation [14]. The difference between the general and the modified Viterbi algorithms is that paths in the trellis (Fig. 1(b)) follow the state transition diagram in Fig. 1(a) with the state transition probabilities in Eqs. (3)–(6). Equation (10) and Fig. 1 establish the principle of the modified Viterbi algorithm.

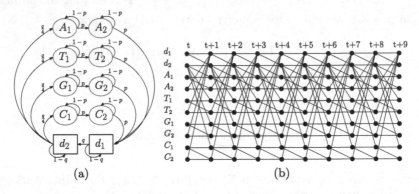

Fig. 1. (a) The state transition diagram of the Markov chain $\{L(t), t \geq 1\}$. In the example shown, $l = 2$ and $r = 2$. (b) An example of the trellis diagram corresponds with the state transition diagram (a).

4   Simulations

4.1   The Single Molecule Synthesis Process

To demonstrate the similarity of the distribution characteristics of our SMSP model in Eqs. (1) and (2) to that of PacBio, we first look at the experiment of PacBio [2]. We assume that the smallest pulse width and the smallest interpulse duration in the SMSP are approximately 100 ms with the time unit in the integrated circuit being 25 ms. That means $l = 5$ and $r = 5$ in Eqs. (1) and (2). Moreover, the largest pulse width and the largest interpulse duration are restricted to 500 ms and 5 s respectively. As we limit the probability $p(Y_n \geq 21(500\,\text{ms})) \leq 10^{-2}$ and $p(W_n \geq 201(5s)) \leq 10^{-3}$, the probability mass functions are adjusted as in Fig. 2 to $p = 0.48$ and $q = 0.06$. The distributions of the simulation results are resembling that of PacBio in terms of pulse characteristics and trace statistics [2].

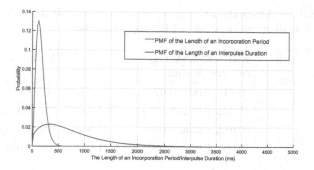

The Length of an Incorporation Period/Interpulse Duration (ms)

Fig. 2. The red line represents the PMF of the length of an incorporation period in Eq. (2) with $p = 0.48$, and the blue line represents the PMF of the length of an interpulse duration in Eq. (1) with $q = 0.06$. (Color figure online)

4.2   The Emission Process and the Received Process

The emission process is to utilize four single-wavelength LEDs (530 nm, 590 nm, 625 nm, and 656 nm) to simulate fluorescence effect on nucleotides (dATP, dTTP, dGTP, and dCTP) where 530 nm LED represents the emission for the phospho-linked dATP and so forth. LED light emission period and time elapsed between light emission are defined as pulse width and interpulse duration in the SMSP model; whereas the pulse width and the interpulse duration are assumed to be in the average of 100 ms and 200 ms and the parameters p and q are estimated from Eqs. (1) and (2) respectively. The intensity of LED light emission is adjusted in the simulation in accordance with a specific series of SNR values. The simulation of the received process involves voltages transformed from light emission intensity and obtained from the photoelectric process, plus Gaussian noise. The SNR is computed as

$$SNR = 10 \cdot \log \frac{P_{signal}}{P_{noise}}.$$

The initial values of the characteristic signals, s_x and $a_x \ \forall \ x \in U$, are measured by a single photodiode from four distinct LEDs and calculated through a simple linear regression in Eq. (8). While the LED light intensity increases as time progresses, the received RGB values are measured, in units of voltage, through photoelectric conversion (PC). The relation between the LED light intensity and the received values results in PC vectors shown in Table 1. The initial value a_d of the ambient light is measured under an LED-free lighting environment. The noise is generated by adding a white Gaussian noise with the mean vector a_d.

Data collection involves the sampling procedures detailed below. The intensity of the LED light emission follows the eight patterns of random walk increments within the following eight ranges: $[30, 60]$, $[40, 80]$, $[50, 100]$, $[60, 120]$, $[70, 140]$, $[80, 160]$, $[90, 180]$, and $[100, 200]$ (photon/ms). For each pattern, 100 samples are generated with each consisting of 1000 random nucleotides. The emission intensity of each nucleotide follows the pattern of random walk , and the durations of the emission depend on the length of the transition state from A

Table 1. The table exhibits the PC vectors measured and calculated from the single photodiode with distinct wavelengths of LED light emission.

Wavelength (nm)	RGB PC Vector s_x (ms·mv/photon)	RGB PC Vector a_x (mv)
530	(0.145, 0.061, 0.024)	(0.903, 1.157, 1.608)
590	(0.093, 0.062, 0.057)	(1.386, 1.161, 1.439)
625	(0.073, 0.056, 0.073)	(1.324, 1.122, 1.401)
656	(0.065, 0.053, 0.085)	(1.073, 0.909, 1.041)

(T, G or C) to d. The corresponding mean SNRs are calculated from these different levels of the light intensities as in Fig. 3. The measured nucleotide sequences are derived through the detection methods in this study. The Smith-Waterman algorithm is then applied to align between the expected sequences and the measured sequences [4,12].

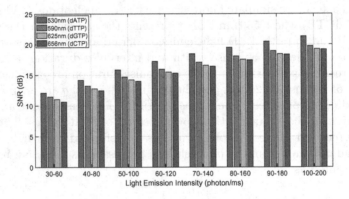

Fig. 3. The relation between the fluctuating light emission intensity (X-axis) and the SNR (Y-axis) of our simulations.

4.3   Results

Deducting from the modeling in above, the sequencing accuracy is beyond 90% even with the fluctuating light emission intensity output ranging from 70 photon/ms to 140 photon/ms and the SNR is below 18 dB. Data analysis is detailed in Fig. 3 and Table 2. In the modeling presented in this study, the accuracy is well beyond 97% even with SNR below 22 dB. Additionally, except mismatches, deletions and insertions of sequencing will not occur under the scenario of SNR beyond 17 dB. The implication signifies that the modeling is outstanding at signal detection even under unfavorable low SNR environment.

Table 2. The table exhibits the results from PacBio platform [2] and our simulations (ML-Viterbi algorithm).

Platform : PacBio	SNR : 22–30 dB			
	A555-dATP	A568-dTTP	A647-dGTP	A660-dCTP
Pulse width (ms)	133 ± 22	91 ± 13	117 ± 14	96 ± 10
Interpulse duration (ms)	770 ± 250	670 ± 220	960 ± 210	790 ± 230
	Correct	Mismatches	Insertions	Deletions
Performance (percentage)	82.9%	4.4%	5.1%	7.6%
ML-Viterbi algorithm	SNR : 16–18 dB			
	530-dATP	590-dTTP	625-dGTP	656-dCTP
Pulse width (ms)	100 ± 5.77	100 ± 5.77	100 ± 5.77	100 ± 5.77
Interpulse duration (ms)	200 ± 18.26	200 ± 18.26	200 ± 18.26	200 ± 18.26
	Correct	Mismatches	Insertions	Deletions
Performance (percentage)	91.57%	8.43%	0%	0%
ML-Viterbi algorithm	SNR : 19–22 dB			
	530-dATP	590-dTTP	625-dGTP	656-dCTP
Pulse width (ms)	100 ± 5.77	100 ± 5.77	100 ± 5.77	100 ± 5.77
Interpulse duration (ms)	200 ± 18.26	200 ± 18.26	200 ± 18.26	200 ± 18.26
	Correct	Mismatches	Insertions	Deletions
Performance (percentage)	97.29%	2.71%	0%	0%

5 Discussion and Conclusions

In contrast to current image processing technology (multiple pixel inputs from CCD camera), single photodiode detection improves processing speed due to the single pixel processing, assuming all other conditions of the two detection methods are identical. Again, the fluorescence labeling technique with single photodiode outperforms conventional single molecule sequencing, especially in the DNA sequencing, with its great accuracy (lower error rate), possible higher throughput, and potentially lower cost. In a combination with the consensus sequencing method, the data accuracy rate can be further improved to nearly flawless [5]. However, the throughput level is highly dependent on how advanced technology has to offer simultaneous sequencing on single photodiode groups. Furthermore, detection accuracy will be greatly improved by augmenting the difference between the fluorescence characteristics for labeling the four kinds of dNTPs. In short, the modeling is definitely an excellent aid to the development of fluorescence based sequencing.

References

1. Ardui, S., Ameur, A., Vermeesch, J.R., Hestand, M.S.: Single molecule real-time (SMRT) sequencing comes of age: applications and utilities for medical diagnostics. Nucleic Acids Res. **46**(5), 2159–2168 (2018). https://doi.org/10.1093/nar/gky066
2. Eid, J., et al.: Real-time DNA sequencing from single polymerase molecules. Science **323**(5910), 133–138 (2009)
3. Epstein, J.R., Biran, I., Walt, D.R.: Fluorescence-based nucleic acid detection and microarrays. Anal. Chim. Acta **469**(1), 3–36 (2002)
4. Gotoh, O.: An improved algorithm for matching biological sequences. J. Mol. Biol. **162**(3), 705–708 (1982)
5. Hiatt, J.B., Patwardhan, R.P., Turner, E.H., Lee, C., Shendure, J.: Parallel, tag-directed assembly of locally derived short sequence reads. Nat. Methods **7**(2), 119 (2010)
6. Horne, K.: An optimal extraction algorithm for ccd spectroscopy. Publ. Astron. Soc. Pac. **98**(604), 609 (1986)
7. Pollard, M.O., Gurdasani, D., Mentzer, A.J., Porter, T., Sandhu, M.S.: Long reads: their purpose and place. Hum. Mol. Genet. **27**(R2), R234–R241 (2018)
8. Roberts, R.J., Carneiro, M.O., Schatz, M.C.: The advantages of SMRT sequencing. Genome Biol. **14**(6), 405 (2013)
9. Schmitt, M.W., Kennedy, S.R., Salk, J.J., Fox, E.J., Hiatt, J.B., Loeb, L.A.: Detection of ultra-rare mutations by next-generation sequencing. Proc. Nat. Acad. Sci. **109**(36), 14508–14513 (2012)
10. Seo, T.S., et al.: Four-color dna sequencing by synthesis on a chip using photocleavable fluorescent nucleotides. Proc. Nat. Acad. Sci. **102**(17), 5926–5931 (2005)
11. Shendure, J., et al.: DNA sequencing at 40: past, present and future. Nature **550**(7676), 345 (2017)
12. Slater, G.S.C., Birney, E.: Automated generation of heuristics for biological sequence comparison. BMC Bioinform. **6**(1), 31 (2005)
13. Valm, A.M., Oldenbourg, R., Borisy, G.G.: Multiplexed spectral imaging of 120 different fluorescent labels. PLoS One **11**(7), e0158495 (2016)
14. Viterbi, A.: Error bounds for convolutional codes and an asymptotically optimum decoding algorithm. IEEE Trans. Inf. Theory **13**(2), 260–269 (1967)
15. van Vuuren, R.D.J., Armin, A., Pandey, A.K., Burn, P.L., Meredith, P.: Organic photodiodes: the future of full color detection and image sensing. Adv. Mater. **28**(24), 4766–802 (2016)

Deep Learning of CTCF-Mediated Chromatin Loops in 3D Genome Organization

Shuzhen Kuang[1] and Liangjiang Wang[2]([⊠])

[1] Department of Biological Sciences, Clemson University, Clemson, SC 29634, USA
[2] Department of Genetics and Biochemistry, Clemson University, Clemson, SC 29634, USA
liangjw@clemson.edu

Abstract. The three-dimensional organization of the human genome is of crucial importance for gene regulation. Results from high-throughput chromosome conformation capture techniques show that the CCCTC-binding factor (CTCF) plays an important role in chromatin interactions, and CTCF-mediated chromatin loops mostly occur between convergent CTCF-binding sites. However, it is still unclear whether and what sequence patterns in addition to the convergent CTCF motifs contribute to the formation of chromatin loops. To discover the complex sequence patterns for chromatin loop formation, we have developed a deep learning model, called DeepCTCFLoop, to predict whether a chromatin loop can be formed between a pair of convergent CTCF motifs using only the DNA sequences of the motifs and their flanking regions. Our results suggest that DeepCTCFLoop can accurately distinguish the convergent CTCF motif pairs forming chromatin loops from the ones not forming loops. It significantly outperforms CTCF-MP, a machine learning model based on word2vec and boosted trees, when using DNA sequences only. Moreover, we show that DNA motifs binding to ASCL1, SP2 and ZNF384 may facilitate the formation of chromatin loops in addition to convergent CTCF motifs. To our knowledge, this is the first published study of using deep learning techniques to discover the sequence motif patterns underlying CTCF-mediated chromatin loop formation. Our results provide useful information for understanding the mechanism of 3D genome organization. The source code and datasets used in this study for model construction are freely available at https://github.com/BioDataLearning/DeepCTCFLoop.

Keywords: Deep learning · CTCF · Sequence motifs · Chromatin loops · 3D genome

1 Introduction

The human genome with more than three billion base pairs is hierarchically organized into three-dimensional (3D) structures to fit into the nucleus. The spatial organization of the genome is critical for the transcriptional control of gene expression and is often disrupted in disease conditions [1, 2]. To characterize the 3D genome architecture, high-throughput techniques have been developed, including the chromosome conformation capture (Hi-C) for detecting global chromatin interactions [3] and chromatin interaction

© Springer Nature Switzerland AG 2020
I. Măndoiu et al. (Eds.): ICCABS 2019, LNBI 12029, pp. 77–89, 2020.
https://doi.org/10.1007/978-3-030-46165-2_7

analysis by paired-end tag sequencing (ChIA-PET) for capturing genome-wide chromatin interactions mediated by specific protein factors [4]. Data from these techniques have shown that the genome is organized into chromatin loops and topologically associating domains (TADs) at the intermediate scale of the spatial organization [5, 6]. Notably, the CCCTC-binding factor (CTCF) binds to the majority of chromatin loop anchors, and CTCF-mediated chromatin loops mostly occur between convergent CTCF-binding sites [5, 6]. The critical role of CTCF in chromatin folding has been confirmed by the loss of loop structures upon CTCF depletion [7]. The functional significance of the convergent orientation between CTCF-binding sites on long-range chromatin looping has also been examined by inverting the CTCF-binding sites using CRISPR/Cas9 [8].

While the majority of CTCF-mediated chromatin loops contain CTCF motifs in a convergent orientation, many pairs of convergent CTCF-binding sites do not form chromatin loops. It is thus interesting to investigate whether and what sequence features in addition to the convergent CTCF motifs are important for establishing CTCF-mediated chromatin loops.

To address this question, a machine learning model, called CTCF-MP, was developed using word2vec and boosted trees [9]. Word2vec is a widely used and computationally efficient method to learn representations of words (word embeddings) using neural networks [10]. It can encode each word in a text corpus as a vector in a continuous vector space where semantically similar words are located near each other. For CTCF-MP, word2vec was used to encode words (k-mers) in the DNA sequences to reduce the dimensionality of words and learn the sequence-based features. It was shown that good model performance was achieved using the word2vec features alone, suggesting the capability of word2vec to capture informative features from the input sequences. However, because the word2vec features are difficult to interpret, the complex sequence pattern underlying CTCF-mediated loop formation remains unclear.

Recently, convolutional neural networks (CNNs), which are designed to process data in multi-dimensional arrays and most commonly applied to computer vision [11], have attracted much attention in the field of biology because of the capability to discover informative sequence motifs directly from input DNA/RNA sequences [12–14]. Besides CNNs, several other advanced deep learning techniques, such as long short-term memory networks (LSTMs) for being able to learn long-range dependencies within sequences [15] and attention mechanisms for being able to capture and emphasize the most important features from the sequential inputs [16], have also been applied to many biological problems [12, 13, 17]. However, these deep learning techniques have not yet been used to discover the complex sequence pattern for the formation of CTCF-mediated chromatin loops.

In this study, we have developed a deep learning model, called DeepCTCFLoop, to predict whether a chromatin loop can be formed between a pair of convergent CTCF motifs and to learn the sequence pattern hidden in the adjacent sequences of CTCF motifs. DeepCTCFLoop utilizes a two-layer CNN and an attention-based bi-directional LSTM (BLSTM) network to learn and emphasize the relevant features, including the sequence motifs, the high-level interactions between sequence motifs, and the long-range dependencies between high-level features. We show that DeepCTCFLoop can accurately predict the chromatin loop formation of convergent CTCF motif pairs within a cell type as well as across cell types. Moreover, our results suggest that the DNA sequence motifs binding to ASCL1, SP2 and ZNF384 may facilitate the formation of CTCF-mediated loops.

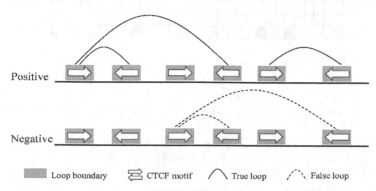

Fig. 1. Schematic diagram of convergent CTCF motifs used to compile the positive and negative instances. The positive instances are defined as the DNA sequences of convergent CTCF motif pairs in the chromatin loop regions and their flanking regions. The negative instances are the DNA sequences of randomly selected convergent CTCF motif pairs not in the chromatin loop regions, under the constraint that the distribution of the distances between the CTCF motif pairs of negative instances is similar as for the positive instances.

2 Materials and Methods

2.1 Data Collection and Preprocessing

DeepCTCFLoop was constructed using data from three different cell types, including GM12878, Hela and K562. To discover the sequence pattern around the convergent CTCF motifs for chromatin loop formation, positive and negative instances were only derived from convergent CTCF motif pairs, as described for CTCF-MP [9]. The positive instances were compiled as the convergent CTCF motif pairs in the chromatin loop regions (Fig. 1) plus 250 nucleotides (nt) on each side of a motif, resulting in a set of DNA sequences of 1038 nt. The locations of CTCF motifs were determined by scanning the human genome (hg19) using FIMO [18] with the known position weight matrix (PWM) of CTCF. The chromatin loop regions of the cell lines GM12878 and Hela were downloaded from NCBI's Gene Expression Omnibus [19] (GEO accession: GSE72816) and the regions of the cell line K562 were obtained from ENCODE [20]. The chromatin loop regions were captured by ChIA-PET for CTCF. To investigate the impact of the flanking region length on the model performance, we also extended the flanking regions to 500 nt, resulting in a slightly worse performance. The negative instances were generated by randomly selecting convergent CTCF motif pairs that were not in the chromatin loop regions, under the constraint that the distribution of the distances between CTCF motif pairs of negative instances was similar as for the positive instances (Fig. 1). The positive and negative instances for each cell type (21,301 positive and

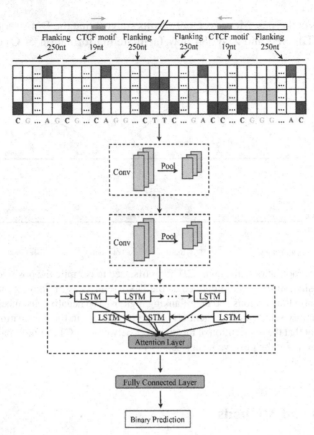

Fig. 2. Diagram of DeepCTCFLoop architecture. The DNA sequence of the convergent CTCF motifs and their surrounding genomic sequences (250 nt) was taken as input by encoding into a binary matrix. Then, a two-layer CNN was adopted to learn the sequence motifs and the high-level interactions between motifs. The BLSTM layer was used to learn the long-range dependencies between the high-level features. Next, an attention layer was used to capture the most important features to increase prediction accuracy. Lastly, two fully connected layers were used to combine the output from the attention layer and make the binary prediction.

21,298 negative instances for GM12878; 9435 and 9432 instances for Hela; and 8205 and 8203 instances for K562) were randomly divided into training, validation and test datasets with the ratio of 80% : 10% : 10%.

2.2 DeepCTCFLoop Model Construction

The architecture of DeepCTCFLoop is shown in Fig. 2. The input of the model is the DNA sequence of convergent CTCF motif pairs and their flanking regions. The input DNA sequence is one-hot-encoded into a 4×1038 binary matrix with A = [1, 0, 0, 0], T = [0, 1, 0, 0], G = [0, 0, 1, 0] and C = [0, 0, 0, 1]. Based on the input, DeepCTCFLoop aims to predict whether a chromatin loop can be formed between a pair of convergent CTCF motifs.

The one-hot-encoded matrix of the DNA input is first fed into a 1D convolution layer. The N filters of the convolution layer with dimension $4 \times L$, where L is the length of a filter, convolve over the input matrix, resulting in N activation maps. The activation value a^s_{fi} for the filter f at the position i of an input sequence s is computed as:

$$a^s_{fi} = \max\left(0, \sum_{l=1}^{L} \sum_{d=1}^{4} w^f_{ld} s_{i+l,d}\right) \tag{1}$$

Here, w^f is the weight matrix for the filter f.

The filters function as motif detectors to discover the patterns within the input sequences. The parameters of the filters can be interpreted as PWMs. High activation values indicate the existence of a motif represented by a PWM at the corresponding positions in the input sequences.

A max pooling layer is used to get the maximum activation value of non-overlapping sub-regions. With the down-sampling strategy, the max pooling layer can reduce input dimensionality and thus avoid model overfitting. Then, a second convolution layer followed by a max pooling layer is used to learn the high-level interactions between the sequence motifs.

Next, a layer of bidirectional long short-term memory network (BLSTM) is used to learn the long-range dependencies among the high-level features learned by the two-layer CNN. LSTM is a variant of the vanilla recurrent neural network (RNN), which scans the input in a sequential manner. Here, BLSTM is used to scan the input both forward and backward. Each LSTM unit consists of an input gate, a forget gate and an output gate. These gates decide what information should be thrown away, be stored, or go to the output [15]. LSTM is thus able to remember the information and learn the long-range dependencies.

Following the BLSTM layer, an attention layer is used to pay more attention on the most important features by assigning more weights to them. The output is then fed into a fully connected layer, and the sigmoid function is used to calculate the probability of forming a chromatin loop.

In this study, the binary cross-entropy loss function was minimized using the Adam optimization algorithm with minibatches [21]. Dropout and L2 regularization were adopted to regularize the model. The early stopping procedure was also used to avoid model overfitting. The model was implemented in Python using Keras 2.2.4 (https://github.com/fchollet/keras) with TensorFlow 1.5.0 as the backend. The hyperparameters of the model were tuned using Bayesian optimization via Hyperopt [22] with the data from GM12878, resulting in the number of CNN filters (N) as 208, the length of filters (L) as 13, the size of pooling layer as 4, the LSTM units as 64, the learning rate as $1e^{-4}$, the L2 regularization as $5e^{-5}$, the dropout rate after CNN as 0.43, and the dropout rate after the attention layer as 0.05. The average time used for model training and evaluation was about 2 h with the data from the three cell types.

2.3 Motif Visualization and Analysis

The filters of the first convolution layer were converted into PWMs as described for Basset [14]. Given a filter f with length L, it scanned all the positive test sequences and calculated an activation value for each position of a sequence. If an activation value was greater than half of the maximum activation m of filter f over all positions of the positive test sequences (Eq. 2), the subsequence corresponding to that activation value was collected. The collected subsequences were aligned and converted into PWMs, which were then visualized using WebLogo [23].

$$m = \max_{s,i} a^s_{fi} \tag{2}$$

The PWMs learned by DeepCTCFLoop were compared with the known motifs in the JASPAR database (2018 vertebrates) [24] using the Tomtom program from the MEME-Suite [25]. Motif pairs with E-value <=0.05 were considered to be significantly matched.

2.4 Model Performance Evaluation

The performance of DeepCTCFLoop was evaluated using the test dataset with the following metrics:

$$\text{Accuracy} = \frac{TP + TN}{TP + TN + FP + FN} \tag{3}$$

$$\text{Sensitivity} = \frac{TP}{TP + FN} \tag{4}$$

$$\text{Specificity} = \frac{TN}{TN + FP} \tag{5}$$

$$\text{MCC} = \frac{TP \times TN - FP \times FN}{\sqrt{(TP + FP)(TP + FN)(TN + FP)(TN + FN)}} \tag{6}$$

Here, TP, TN, FP, and FN represent the number of true positives, true negatives, false positives and false negatives, respectively. Matthews correction coefficient (MCC) is generally considered as a robust measurement of model performance with a balanced or imbalanced dataset. Moreover, the receiver operating characteristic (ROC) curve and the area under the ROC curve (ROC AUC) are also used for model evaluation. The ROC curve and ROC AUC are considered to be more robust for an imbalanced dataset than the other performance metrics.

3 Results and Discussion

DeepCTCFLoop has been developed to predict the chromatin loop formation between convergent CTCF motif pairs, and to discover sequence motif patterns for the loop formation (Fig. 1).

3.1 DeepCTCFLoop Could Accurately Predict Chromatin Loops Formed By Convergent CTCF Motifs

DeepCTCFLoop takes convergent CTCF motif pairs and their surrounding genomic sequences as input to predict chromatin loop formation (Fig. 2). The performance of DeepCTCFLoop was evaluated using datasets derived from three different cell types, including GM12878, Hela and K562. The hyperparameters for model construction were optimized using the dataset of GM12878 (see Materials and Methods). Dropout, L2 regularization and the early stopping procedure were used to avoid model overfitting.

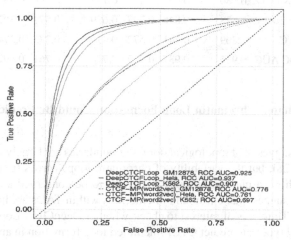

Fig. 3. ROC curves of DeepCTCFLoop and CTCF-MP (word2vec features only) on the test datasets of GM12878, Hela and K562.

As shown in Fig. 3 and Table 1, DeepCTCFLoop achieved the mean ROC AUC of 0.925 for GM12878, 0.937 for Hela and 0.907 for K562 on the test datasets for 10 repetitions. The high model performance indicates that DeepCTCFLoop can learn informative features besides the motif orientation from the DNA sequences to distinguish interacting convergent CTCF motif pairs (positive instances) from non-interacting ones (negative instances). By comparison, CTCF-MP [9] achieved relatively poor performance on the same datasets with the mean ROC AUC of 0.776 for GM12878, 0.761 for Hela and 0.697 for K562, when only using the DNA sequence features from word2vec (Fig. 3 and Table 1). The superior performance of DeepCTCFLoop over CTCF-MP was also suggested by the significantly higher accuracy, sensitivity, specificity and MCC values (Table 1). Although word2vec may capture the contextual information between k-mers (DNA words) by learning their semantical similarity, the results from this study suggest that DeepCTCFLoop can capture more relevant information, such as sequence motifs, from the input DNA sequences. This is consistent with the poor performance of word2vec on detecting informative motifs in a previous study [26]. Taken together, our results demonstrate the superior capability of DeepCTCFLoop to predict the formation of chromatin loops mediated by convergent CTCF motif pairs.

Table 1. Superior performance of DeepCTCFLoop over CTCF-MP when only using DNA sequences as the input. For CTCF-MP, DNA sequences were encoded into vector features by word2vec. The average accuracy, sensitivity, specificity, Matthews correction coefficient (MCC), and the area under the receiver operating characteristic curve (ROC AUC) for 10 repetitions of the two models on the test datasets from GM12878, Hela and K562 cells are shown.

Metrics	DeepCTCFLoop			CTCF-MP (word2vec)		
	GM12878	Hela	K562	GM12878	Hela	K562
Accuracy	0.854	0.870	0.834	0.703	0.682	0.622
Sensitivity	0.880	0.894	0.872	0.775	0.798	0.850
Specificity	0.828	0.846	0.798	0.629	0.571	0.392
MCC	0.709	0.741	0.672	0.410	0.375	0.274
ROC AUC	0.925	0.937	0.907	0.776	0.761	0.697

3.2 CTCF-Mediated Chromatin Loop Formation Could Be Predicted Across Cell Types

Previous studies suggest that topological domain boundaries are largely invariant across tissue types [27, 28], but the variability of chromatin loops across tissue types is still unclear. Although most chromatin loops are found to be conserved among cell types [5], some studies suggest that chromatin interactions within TADs are highly cell-type-specific [29, 30]. Thus, it is of interest to check whether DeepCTCFLoop trained using data from one cell type can predict CTCF-mediated loop formation in another cell type.

To obtain an unbiased performance evaluation, the positive and negative instances shared between two cell types were removed from the test cell type (6806 positive and 3209 negative instances shared between GM12878 and Hela; 5730 and 3378 instances between GM12878 and K562; 4209 and 3133 instances between Hela and K562). As shown in Table 2, DeepCTCFLoop trained with the data from one cell type can accurately predict the chromatin loops in the other two cell types. Moreover, the cross-cell-type performance of DeepCTCFLoop is higher than the same-cell-type performance of CTCF-MP using word2vec features, further suggesting the higher capability of DeepCTCFLoop in capturing the complex sequence pattern for chromatin loop

Table 2. DeepCTCFLoop performance for the cross-cell-type prediction of chromatin loop formation. The average area under the receiver operating characteristic curve (ROC AUC) from 10 repetitions is shown.

Test cell type	Training cell type		
	GM12878	Hela	K562
GM12878	**0.925**	0.814	0.791
Hela	0.852	**0.937**	0.843
K562	0.759	0.766	**0.909**

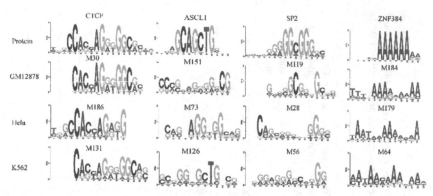

Fig. 4. Sequence logos of the PWMs significantly matched to the DNA motifs of the proteins CTCF, ASCL1, SP2 and ZNF384. The PWMs learned by DeepCTCFLoop from the GM12878, Hela and K562 datasets were compared with the known motifs in the JASPAR database using Tomtom.

formation. Taken together, the high performance of DeepCTCFLoop in cross-cell-type prediction suggests its potential to predict CTCF-mediated chromatin loops in a new cell type.

3.3 Interesting DNA Sequence Motifs Were Discovered by DeepCTCFLoop

The ability of DeepCTCFLoop to distinguish interacting CTCF motif pairs from non-interacting ones suggests that it may have learned the complex sequence pattern for CTCF-mediated loop formation. To understand the sequence pattern, the filters of the first convolutional layer were converted into PWMs as described in Materials and Methods. For the model built with data from GM12878 cells (GM12878 model), 189 PWMs were derived and compared with the known transcription factor (TF) motifs in the JASPAR database [24] using Tomtom [25]. With E-value <=0.05, 34 of the 189 PWMs were significantly matched to the known motifs, including 30 PWMs matched to the motifs of 34 non-CTCF proteins. It is likely that these DNA-binding proteins are involved in the chromatin loop formation mediated by convergent CTCF motif pairs.

To discover the consistent motif pattern that may contribute to CTCF-mediated chromatin loop formation across different cell types, we performed the same motif analysis for the models built using the data from Hela cells (Hela model) and K562 cells (K562 model). Similarly, 34 of the 204 PWMs (Hela model) and 35 of the 208 PWMs (K562 model) were significantly matched to 27 and 21 non-CTCF motifs, respectively. Interestingly, besides CTCF, the DNA motifs of several proteins (ASCL1, SP2 and ZNF384) were commonly matched by the PWMs learned from the three different models (GM12878, Hela and K562 model) (Fig. 4 and Table 3). ASCL1, an evolutionarily conserved basic-helix-loop-helix (bHLH) transcription factor, has been shown to promote local chromatin accessibility at its target regions during neurogenesis and is associated with transcription activation by mostly binding to distal enhancers [31–33]. As CTCF-mediated chromatin loops are involved in promoter-enhancer interactions [34], the binding of ASCL1 to distal enhancers may promote the loop formation. Moreover,

Table 3. List of some interesting PWMs learned by DeepCTCFLoop. These PWMs were learned from the data of three cell types (GM12878, Hela and K562) and were significantly matched to the known DNA motifs of ASCL1, SP2 and ZNF384 by Tomtom. The consensus sequences of the learned PWMs and the known motifs, and genes associated with the known motifs are shown.

DeepCTCFLoop PWM ID			Consensus sequence of DeepCTCFLoop PWM			Consensus sequence of the known motif in JASPAR	Gene associated with the known motif
GM12878	Hela	K562	GM12878	Hela	K562		
M151	M73	M36	CCCGAAGGGGGCG	CAGCAGGTGGCAG	ACCAGCGGGCGCT	GCAGCAGCTGGCG	ASCL1
	M121	M126		GGCCAGCAGGGGG	GCGGGAGCTGCCG		
		M128			ACAGCTGCTGGAG		
M119	M28	M56	GGGGGCGGGGCGG	CAGCGGGGTGGGGG	GGGAGGGGCAGGG	GCAGCAGCTGGCG	SP2
M131	M56	M30	AAAAATTAAAAAA	AATAAATAAAAAT	TTATTTATTTATT	TTTTAAAAAAAAA	ZNF384
M168	M64	M64	TATAAAAAAATAC	TTTTCTTTTTTTA	AATAACAAAAAAA		
M184	M149	M93	TTTTAAAAAAAAA	AATTCGTTTTTTT	AAAAAAAAAAAAA		
M187	M179	M95	TTTGTTTTTTTAAA	TAATAAAAAAAAA	TTAAATAAATTCA		
	M203	M189		AATAGAAAAAAAA	AAAAAAAAAAAAG		
		M196			TTTTTTTCTAGTT		

ZNF384, a C2H2-type zinc finger protein, has been shown to be directly involved in chromatin looping, like CTCF and the cohesin complex. The interaction of ZNF384 with CTCF may contribute to the sequence specificity of the chromatin loop formation [35]. Although SP2, a member of the SP family with a conserved DNA-binding domain, has not been reported to be involved in chromatin loop formation, it mainly localizes to subnuclear foci associated with the nuclear matrix and is mostly involved in gene activation [36]. The results suggest that the binding of these proteins to specific DNA motifs may provide additional information for the chromatin loop formation mediated by convergent CTCF motif pairs.

4 Conclusion

In this study, we have developed a deep learning model, called DeepCTCFLoop, to predict whether a chromatin loop can be formed between a pair of CTCF motifs in the convergent orientation, and to discover the sequence pattern besides the convergent CTCF motifs for the loop formation. The CTCF motif pairs and their flanking genomic sequences were used as model input. Through evaluating on three different cell types (GM12878, Hela and K562), DeepCTCFLoop was demonstrated to be able to accurately predict CTCF-mediated chromatin loop formation. It significantly outperformed a previous machine learning model (CTCF-MP with word2vec features only). DeepCTCFLoop also showed high performance across cell types. Interestingly, the DNA motifs of several proteins (ASCL1, SP2 and ZNF384) were significantly matched with the PWMs learned by DeepCTCFLoop from the data of GM12878, Hela and K562 cells, suggesting the potential roles of these proteins in CTCF-mediated loop formation. In particular, ZNF384 has been reported to be involved in chromatin loop formation [35]. Previous studies suggest that mutations in the boundaries of chromatin loops could cause loop alterations, leading to disrupted gene expression [37, 38]. Recently, the increasing applications of whole-genome sequencing to various diseases lead to the identification of many disease-associated non-coding variants, most of which have unknown mechanisms. DeepCTCFLoop can be applied to these non-coding variants to predict their effects on the formation of CTCF-mediated chromatin loops, and thus help annotate these non-coding variants.

References

1. Bonev, B., Cavalli, G.: Organization and function of the 3D genome. Nat. Rev. Genet. **17**, 661 (2016)
2. Bickmore, W.A.: The spatial organization of the human genome. Ann. Rev. Genomics Hum. Genet. **14**, 67–84 (2013)
3. Lieberman-Aiden, E., Van Berkum, N.L., Williams, L., Imakaev, M., Ragoczy, T., Telling, A., et al.: Comprehensive mapping of long-range interactions reveals folding principles of the human genome. Science **326**, 289–293 (2009)
4. Fullwood, M.J., Liu, M.H., Pan, Y.F., Liu, J., Xu, H., Mohamed, Y.B., et al.: An oestrogen-receptor-α-bound human chromatin interactome. Nature **462**, 58 (2009)

5. Rao, S.S., Huntley, M.H., Durand, N.C., Stamenova, E.K., Bochkov, I.D., Robinson, J.T., et al.: A 3D map of the human genome at kilobase resolution reveals principles of chromatin looping. Cell **159**, 1665–1680 (2014)
6. Tang, Z., Luo, O.J., Li, X., Zheng, M., Zhu, J.J., Szalaj, P., et al.: CTCF-mediated human 3D genome architecture reveals chromatin topology for transcription. Cell **163**, 1611–1627 (2015)
7. Nora, E.P., Goloborodko, A., Valton, A.-L., Gibcus, J.H., Uebersohn, A., Abdennur, N., et al.: Targeted degradation of CTCF decouples local insulation of chromosome domains from genomic compartmentalization. Cell **169**, 930–944 (2017). e922
8. Guo, Y., Xu, Q., Canzio, D., Shou, J., Li, J., Gorkin, D.U., et al.: CRISPR inversion of CTCF sites alters genome topology and enhancer/promoter function. Cell **162**, 900–910 (2015)
9. Zhang, R., Wang, Y., Yang, Y., Zhang, Y., Ma, J.: Predicting CTCF-mediated chromatin loops using CTCF-MP. Bioinformatics **34**, i133–i141 (2018)
10. Mikolov, T., Chen, K., Corrado, G., Dean, J.: Efficient estimation of word representations in vector space (2013). arXiv preprint arXiv:1301.3781
11. LeCun, Y., Bengio, Y., Hinton, G.: Deep learning. Nature **521**, 436 (2015)
12. Quang, D., Xie, X.: DanQ: a hybrid convolutional and recurrent deep neural network for quantifying the function of DNA sequences. Nucleic Acids Res. **44**, e107 (2016)
13. Angermueller, C., Lee, H.J., Reik, W., Stegle, O.: DeepCpG: accurate prediction of single-cell DNA methylation states using deep learning. Genome Biol. **18**, 67 (2017)
14. Kelley, D.R., Snoek, J., Rinn, J.L.: Basset: learning the regulatory code of the accessible genome with deep convolutional neural networks. Genome Res. **26**, 990–999 (2016)
15. Hochreiter, S., Schmidhuber, J.: Long short-term memory. Neural Comput. **9**, 1735–1780 (1997)
16. Zhou, P., Shi, W., Tian, J., Qi, Z., Li, B., Hao, H., et al.: Attention-based bidirectional long short-term memory networks for relation classification. In: Proceedings of the 54th Annual Meeting of the Association for Computational Linguistics (Volume 2: Short Papers), pp. 207–212 (2016)
17. Li, W., Wong, W.H., Jiang, R.: DeepTACT: predicting 3D chromatin contacts via bootstrapping deep learning. Nucleic Acids Res. **47**, e60–e60 (2019)
18. Grant, C.E., Bailey, T.L., Noble, W.S.: FIMO: scanning for occurrences of a given motif. Bioinformatics **27**, 1017–1018 (2011)
19. Edgar, R., Domrachev, M., Lash, A.E.: Gene expression omnibus: NCBI gene expression and hybridization array data repository. Nucleic Acids Res. **30**, 207–210 (2002)
20. Consortium, E.P.: The ENCODE (ENCyclopedia of DNA elements) project. Science **306**, 636–640 (2004)
21. Kingma, D.P., Ba, J.: Adam: A method for stochastic optimization (2014). arXiv preprint arXiv:1412.6980
22. Bergstra, J., Yamins, D., Cox, D.D.: Hyperopt: A python library for optimizing the hyper-parameters of machine learning algorithms. Proceedings of the 12th Python in Science Conference, pp. 13-20 (2013)
23. Crooks, G.E., Hon, G., Chandonia, J.-M., Brenner, S.E.: WebLogo: a sequence logo generator. Genome Res. **14**, 1188–1190 (2004)
24. Mathelier, A., Zhao, X., Zhang, A.W., Parcy, F., Worsley-Hunt, R., Arenillas, D.J., et al.: JASPAR 2014: an extensively expanded and updated open-access database of transcription factor binding profiles. Nucleic Acids Res. **42**, D142–D147 (2013)
25. Bailey, T.L., Boden, M., Buske, F.A., Frith, M., Grant, C.E., Clementi, L., et al.: MEME SUITE: tools for motif discovery and searching. Nucleic Acids Res. **37**, W202–W208 (2009)
26. Trabelsi, A., Chaabane, M., Hur, A.B.: Comprehensive Evaluation of Deep Learning Architectures for Prediction of DNA/RNA Sequence Binding Specificities (2019). arXiv preprint arXiv:1901.10526

27. Dixon, J.R., Selvaraj, S., Yue, F., Kim, A., Li, Y., Shen, Y., et al.: Topological domains in mammalian genomes identified by analysis of chromatin interactions. Nature **485**, 376 (2012)
28. Dekker, J., Heard, E.: Structural and functional diversity of topologically associating domains. FEBS Lett. **589**, 2877–2884 (2015)
29. Smith, E.M., Lajoie, B.R., Jain, G., Dekker, J.: Invariant TAD boundaries constrain cell-type-specific looping interactions between promoters and distal elements around the CFTR locus. Am. J. Hum. Genet. **98**, 185–201 (2016)
30. Bouwman, B.A., de Laat, W.: Getting the genome in shape: the formation of loops, domains and compartments. Genome Biol. **16**, 154 (2015)
31. Aydin, B., Kakumanu, A., Rossillo, M., Moreno-Estellés, M., Garipler, G., Ringstad, N., et al.: Proneural factors Ascl1 and Neurog2 contribute to neuronal subtype identities by establishing distinct chromatin landscapes. Nat. Neurosci. **22**(6), 897–908 (2019)
32. Raposo, A.A., Vasconcelos, F.F., Drechsel, D., Marie, C., Johnston, C., Dolle, D., et al.: Ascl1 coordinately regulates gene expression and the chromatin landscape during neurogenesis. Cell Rep. **10**, 1544–1556 (2015)
33. Park, N.I., Guilhamon, P., Desai, K., McAdam, R.F., Langille, E., O'Connor, M., et al.: ASCL1 reorganizes chromatin to direct neuronal fate and suppress tumorigenicity of glioblastoma stem cells. Cell Stem Cell **21**, 209–224 (2017). e207
34. Ren, G., Jin, W., Cui, K., Rodrigez, J., Hu, G., Zhang, Z., et al.: CTCF-mediated enhancer-promoter interaction is a critical regulator of cell-to-cell variation of gene expression. Mol. Cell **67**, 1049–1058 (2017). e1046
35. Whalen, S., Truty, R.M., Pollard, K.S.: Enhancer–promoter interactions are encoded by complex genomic signatures on looping chromatin. Nat. Genet. **48**, 488 (2016)
36. Moorefield, K.S., Yin, H., Nichols, T.D., Cathcart, C., Simmons, S.O., Horowitz, J.M.: Sp2 localizes to subnuclear foci associated with the nuclear matrix. Mol. Biol. Cell **17**, 1711–1722 (2006)
37. Hnisz, D., Weintraub, A.S., Day, D.S., Valton, A.-L., Bak, R.O., Li, C.H., et al.: Activation of proto-oncogenes by disruption of chromosome neighborhoods. Science **351**, 1454–1458 (2016)
38. Guo, Y.A., Chang, M.M., Huang, W., Ooi, W.F., Xing, M., Tan, P., et al.: Mutation hotspots at CTCF binding sites coupled to chromosomal instability in gastrointestinal cancers. Nat. Commun. **9**, 1520 (2018)

Effects of Various Alpha-1 Antitrypsin Supplement Dosages on the Lung Microbiome and Metabolome

Trevor Cickovski[1]([✉]), Astrid Manuel[2], Kalai Mathee[3], Michael Campos[4], and Giri Narasimhan[1]

[1] Bioinformatics Research Group (BioRG), Florida International University, Miami, FL, USA
{tcickovs,giri}@fiu.edu
[2] Center for Precision Health, University of Texas Health Science Center, Houston, TX, USA
Astrid.M.Manuel@uth.tmc.edu
[3] Human and Molecular Genetics, Florida International University, Miami, FL, USA
Kalai.Mathee@fiu.edu
[4] Miller School of Medicine, University of Miami, Miami, FL, USA
mcampos1@med.miami.edu

Abstract. Patients with Alpha-1 Antitrypsin Deficiency (A1AD) have abnormally low levels of the protein *Alpha-1 Antitrypsin* (AAT) in their blood, because of a double mutation that makes the protein misfold and instead collect in the liver (sometimes even causing cirrhosis). The currently accepted single dosage (SD) of AAT supplements does not produce AAT blood concentrations anywhere near normal levels; they typically only reach the effect of having a single mutation. Some have therefore advocated for a double dosage (DD) of these treatments, which generally would be enough to approach these normal concentrations. Levels of *cytokines*, produced by the immune system in response to an attack, have already been observed to drop dramatically when A1AD patients consuming single dosage started taking double dosage, and then either remain the same or increase again upon return to a single dosage regimen. In this study we administer the same dosage sequence to A1AD patients (SD, DD, SD) for one month each and view the effects on their lung *microbiome* and *metabolome*. We analyze both at the end of each stage, comparing and contrasting and discovering potential biomarkers for each stage, and concluding with a discussion of potential implications.

Keywords: Microbiome · Metabolome · Networks · Alpha-1 Antitrypsin Deficiency (A1AD)

1 Introduction

Abnormally low blood concentrations of the protein Alpha-1 Antitrypsin (AAT) produce a condition known as Alpha-1 Antitrypsin Deficiency (A1AD, [26]).

© Springer Nature Switzerland AG 2020
I. Măndoiu et al. (Eds.): ICCABS 2019, LNBI 12029, pp. 90–101, 2020.
https://doi.org/10.1007/978-3-030-46165-2_8

AAT is a multi-function protein, modulating immunity, inflammation, proteostasis, apoptosis, and cellular senescence [19]. These are all critical for lung maintenance, particularly against harmful substances (i.e., cigarette smoke). AAT is produced in the liver, and transferred to the lung through the bloodstream. However, if the gene that produces AAT becomes mutated the protein misfolds, causing it to remain in the liver instead of entering the bloodstream. In addition to potentially leading to cirrhosis [38], this misfolding produces abnormally low levels of AAT in the bloodstream, resulting in not enough AAT being transferred to the lungs. Normal levels of blood AAT concentration are above 25 μM. A1AD *carriers* have a single mutation in the AAT gene, causing blood levels to drop to around 15 μM. Double mutations in this gene yield an A1AD diagnosis, producing AAT blood concentrations as dangerously low as 4 μM.

Treatments for lung diseases resulting from A1AD (i.e. emphysema or COPD) involve augmentation or *replacement* therapy [18], where patients are given AAT supplements from healthy donor blood plasma. The medically adopted standard single dosage (SD), 60 mg/kg/week [41], will generally achieve a target AAT blood concentration of about 11 μM. This is slightly under carrier-level concentrations. Recent studies have questioned this dosage, with some advocating for a double-dosage (DD) that produces AAT blood concentrations that approach normal levels [6]. One in particular showed reduced cytokine (produced by the immune system in response to an infection or attack) levels when patients switched from SD to DD, many of which began increasing again upon return to a SD [5]. These elevated cytokine levels (implying increased immune system activity) for SD patients before and after their DD period call into question whether or not the SD yields enough AAT lung transfer to maintain healthy functionality.

The *microbiome* has recently gained attention because the number of microbial cells in a human body are estimated to exceed that of the number of human cells [36], and therefore hold enormous potential to contribute to our health, both positively and negatively. Compared to the more heavily studied gut microbiome the lung microbiome has a significantly lower biomass but despite this, generally exhibits high diversity in its community [27]. This diversity has been shown to be lower in the airway microbiome of asthma patients, and its composition can even be used as a biomarker to indicate severity [10]. Pathogenic bacteria from genera such as *Pseudomonas* and *Staphylococcus* have been implicated in Cystic Fibrosis [42], and inhaled antibiotics have long been used as treatments [9].

Significance of the microbiome is largely because of metabolites produced by member microbiota that become involved in underlying chemical reactions with metabolites produced by other microbiota and the host, collectively influencing host functionality. The collection of metabolites in a host or environment is termed the *metabolome*, and this can help to explain the mechanisms behind the influence of the microbiome on human health. It would be interesting therefore to see if performing the same augmentation therapy that we mentioned (SD, followed by DD, followed by SD) results in similar shifts in the microbiome and metabolome to those observed in the cytokines. Additionally, such analyses would increase our general depth of knowledge regarding the effects of both dosages on important underlying biological circuitry.

We collect metagenomics and metabolomics samples from A1AD patients on an augmentation therapy plan of one month SD, followed by one month DD, and finally one month SD. Our collection takes place at the end of every month. Our analysis then compares and contrasts all pairs of sample sets from the perspectives of diversity, composition and ecological relationships, and produces lists of distinguishing microbes and metabolites for each sample set. We conclude by discussing results, potential biological implications, and future directions.

2   Methods

We collected lung microbiome samples from ten subjects diagnosed with A1AD, over a three-month clinical trial period. For the first month these patients were administered the drug *Zemaira* [3] at a standard, FDA-approved dosage of 60 mg/kg/week (*SD1*). They then were administered a double dose (120 mg/kg/week, *DD*) for the second month, followed by one final month back at the single (*SD2*). Samples were collected at the end of each of period, using a bronchoscopy involving lung lavage, brushings and endobronchial biopsies. We then performed metagenomics and metabolomics analyses on these samples.

Metagenomics. Zymo Research Metagenomic Services (Orange, CA) performed our 16S sequencing, using standard protocols [2,13,17,20,24,29,40]. The final set of amplicon sequences were compiled, clustered and then analyzed using Qiime [7]. When classifying taxa, we used the SILVA [31] reference database and removed singleton taxa. We then used Qiime for other downstream metagenomics analysis, including alpha- and beta-diversity plots of the microbial communities compared across all three sample sets. We also used Qiime to compare relative abundances of each OTU in the three sample sets. Finally, we built Microbial Co-occurence Networks (MCNs) [14] to analyze ecological relationships between taxa within the microbiome.

Metabolomics. For metabolomics analysis we used several programs in the MetaboAnalyst [44] suite of tools after normalizing our concentration data. We produce distinguishing metabolites for each pair of sets using both volcano plots and Partial Least Squares Differential Analysis (PLS-DA, [1]), and test agreement between the metabolites they generate. PLS-DA has also become a standard classification tool in metagenomics, but it has some weaknesses [32] particularly with respect to over-fitting and a lack of cross-validation. Therefore we also attempt to classify samples using a Random Forest [4], where agreement between multiple decision tree learners can provide an indication of the quality of dissimilarity between sample sets. We then produce another set of distinguishing metabolites used by Random Forest when classifying sample sets.

3   Results and Discussion

3.1   Metagenomics

While beta-diversity results were inconclusive, alpha-diversity results exhibited agreement across the four standard algorithms offered by Qiime as shown in

Fig. 1: (a) The *Chao* metric [8] which measures richness, (b) *Observed species*, or unique taxa counts in each sample, (c) *PD_whole_tree* which measures phylogenetic distance between community members, and (d) the *Shannon index* [37]. The plots show *SD1* (red), *DD* (blue), and *SD2* (orange). Although error bars were large enough to encompass all three curves, there is strong agreement between all four plots on the average relative tracks of these curves, with *DD* and *SD2* close in magnitude and *SD1* noticeably lower.

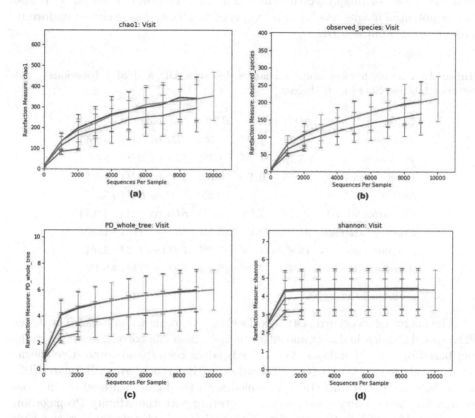

Fig. 1. Alpha diversity for lung microbiome samples collected at *SD1* (red), *DD* (blue) and *SD2* (orange) using (a) Chao richness, (b) observed species (unique taxa), (c) phylogenetic distance, and (d) Shannon index. (Color figure online)

To view taxa most affected by the DD (*DD*), we extracted those with a relative abundance change magnitude larger than 1% when comparing *SD1* to *DD*, and repeated the process to compare *SD2* to *DD*. Table 1 summarizes our results, with the "Notable Change" column designating the two augmentation therapy stages being compared. There are two trends in this table of which we take note. First, all taxa that *decreased* in relative abundance from *SD1* to *DD* tended to remain at that same lower abundance when moving to *SD2*. On the other hand, taxa that significantly *increased* in relative abundance from *SD1* to

DD tended to return to the lower *SD1* levels in *SD2*. This is an interesting parallel with the cytokine behavior, which was also observed to either return to *SD1* levels or remain the same in *SD2*. Secondly, we note a sharp (collectively over 20%) drop in the relative abundance of two highly abundant taxa (*Propionibacterium* and *Streptococcus*) when advancing from *DD* to *SD2*. About 8.14% of the increase came from the genus *Actinomyces*, and the remainder came from genera *Pelomonas*, *Acinetobacterium*, *Cardiobacterium*, *Brevundimonas*, and *Neisseria* from the class Gammaproteobacterium. The last three were confired by LEfSe [35] as potential biomarkers for *SD2* compared to *DD*, and were almost uniformly present in *SD2* but not *DD*.

Table 1. Taxa with most notable changes between DD and SD1 (previous), and between DD and SD2 (next) stages.

Taxon	SD1	DD	SD2	Notable change
Prevotella	7.0%	1.0%	1.6%	*SD1* to *DD* (−5.9%)
Veillonella	3.3%	1.2%	1.7%	*SD1* to *DD* (−2.1%)
Parvimonas	1.7%	0.4%	0.8%	*SD1* to *DD* (−1.3%)
Rothia	2.3%	7.6%	2.2%	*SD1* to *DD* (+5.3%)
Corynebacterium	2.2%	5.1%	2.1%	*SD1* to *DD* (+2.9%)
Propionibacterium	31.4%	32.1%	15.5%	*DD* to *SD2* (−16.6%)
Streptococcus	12.8%	11.2%	7.3%	*DD* to *SD2* (−3.9%)
Actinomyces	7.9%	7.9%	16.1%	*DD* to *SD2* (+8.1%)

The microbial co-occurrence networks (MCNs) are shown in Fig. 2. The nodes correspond to microbial taxa and edges to Spearman [39] correlations with ranking based on taxa abundance. Green (red) edges correspond to positive (negative) correlations. The MCNs are visualized using Fruchterman-Reingold [16], with the eight nodes from Table 1 labeled with the taxon centered on the corresponding node. Finally, we perform clustering with the Affinity Propagation algorithm [15] which works with signed networks, and color the clusters to which the above nodes have connections. For clarity we remove all nodes with no edges, except *Corynebacterium* in *SD1* as it was one of the taxa from Table 1.

We note that the three taxa that dropped in abundance when going from *SD1* to *DD* (*Prevotella*, *Veillonella*, and *Parvimonas*) all occupied "centralized" locations in the *SD1* network and contributed positively to multiple clusters (collectively 44 edges spanning five different clusters). Although most edges were still positive in the *DD* network, there were far fewer (29) and some were negative (to *Acinetobacteria*, *Pelomonas* and *Ralstonia*). Returning to a single dose (*SD2*) had mixed effects. While it did increase the collective number of edges back into the 40s and also the number of positive connections, negatives remained that were not present in *SD1*. The two taxa that increased in abundance when going

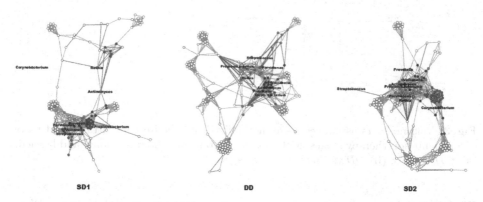

Fig. 2. Microbial Co-occurence Networks (MCNs) for each augmentation therapy period. SD1 = first single dose, DD = double dose, SD2 = second single dose (Color figure online)

from *SD1* to *DD* (*Rothia* and *Corynebacterium*) seemed to become more "integrated", suggesting an increased interaction with other taxa during this transition. In particular, *Corynebacterium*, which was completely disconnected from the *SD1* network, was part of a cluster in the *DD* network with 11 positive edges and connections to three other clusters. In *SD2* that number grew to 23 edges, all positive, to two larger and tightly connected clusters (particularly the red). Note that *Corynebacterium* continued to become more integrated even as its abundance dropped when moving from *DD* to *SD2*. *Rothia* seemed to experience something similar, though was connected more negatively in *SD1* (2 positive edges, 4 negative) but became more integrated with time (5 positive edges and 2 negative in *DD*, 10 positive and 1 negative in *SD2*). The remaining three taxa that experienced large fluctuations from *DD* to *SD2* (*Propionibacterium*, *Streptococcus* and *Actinomyces*), also showed some changes here. The *SD1* shows more "polarized" connections for the first two, with *Propionibacterium* having all negative edges and *Streptococcus* all positive. This changes in *DD* and *SD2*, particularly with *Streptococcus* in *SD2* which is almost completely "pushed out" of the network with its sole connection a strongly negative correlation to *Anaerococcus*. Interestingly, *Anaerococcus* is also the only positively correlated taxa to *Propionibacterium* in this same network. The effect of *Actinomyces* also seems to rise, culminating in ten edges (all positive) to five different clusters in *SD2*.

3.2 Metabolomics

Volcano Plots. Figure 3 shows volcano plots comparing each pair of stages: ((a) *SD1* and *DD*, (b) *DD* and *SD2*, and (c) *SD1* and *SD2*). On each plot, the x-axis represents the log fold-change between sample set pairs and the y-axis the negative log of the p-value. Distinguishing metabolites have high fold-change and low p-value, and these are found in the upper left and right corners. The most distinguishing metabolites are represented as pink dots, labelled with the corresponding metabolite name.

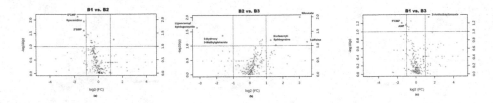

Fig. 3. Volcano plots assembled using normalized metabolite concentrations of pairs of augmentation therapy stages, with most important metabolites colored and labeled. (a) *SD1* vs. *DD*; (b) *DD* vs. *SD2*; (c) *SD1* vs. *SD2*.

PLS-DA. Additionally we ran Partial Least Squares Differential Analysis (PLS-DA, [1]) a total of three times, once for each pair of augmentation therapy stages. This produced a list of the most distinguishing metabolites between each pair of stages. Table 2 shows these metabolites, the pair of stages they distinguish, and the stage in which they were elevated in **bold**. It can be noted that key metabolites discovered by the volcano plot and PLS-DA exhibit high level agreement, offering mutual support. We also note a tendency for these metabolites to be elevated in *DD* compared to other stages, and in *SD2* compared to *SD1*. This is significant because both *DD* and *SD2* take place after the *SD1*. Table 3 shows this tendency to continue across all distinguishing metabolites.

Table 2. Distinguishing metabolites when running PLS-DA.

Metabolite	Distinguishes (**Elevated**)
5'CMP	SD1/**DD**
Spermidine	SD1/**DD**
3'GMP	SD1/**DD**
Cholesterol	SD1/**DD**
2-Aminoheptanoate	**SD1**/DD
Ribonate	**DD**/SD2
Lignoceroyl Sphingomyerlin	DD/**SD2**
3-Hydroxy 3-Methylglutarate	DD/**SD2**
N-Stearyl Shingosine	**DD**/SD2
Caffeine	**DD**/SD2
1-(1-Enyl-Palmitoyl)-2-Oleoyl-GPE	**DD**/SD2
2-Aminoheptanoate	**SD1**/SD2
5'CMP	SD1/**SD2**
AMP	SD1/**SD2**

Random Forest. For Random Forest we performed a total of six runs, since Random Forest attempts to classify samples as belonging to a specific set, rather

Table 3. Comparing concentration levels of 14 distinguishing PLS-DA metabolites between pairs of augmentation therapy stages.

Comparison	Elevated in *SD1*	Elevated in *DD*	Elevated in *SD2*
SD1 vs. *DD*	1	13	
DD vs. *SD2*		12	2
SD1 vs. *SD2*	3		11

than distinguishing between two sets (therefore *SD1* vs. *DD* and *DD* vs. *SD1* are now different). Results were mixed, but Random Forest exhibited the most success when classifying *SD1* samples (70% and 60% accuracy respectively alongside *DD* and *SD2* samples). The top ten metabolites identified by Random Forest when distinguishing *SD1* vs. *DD* and *SD1* vs. *SD2* are shown in Table 4. In each column we denote the later stage along with elevated metabolites in that stage in **bold**. We note the continuing trend of larger concentrations of significant metabolites after the *DD* had been administered. All ten metabolites were elevated in *DD* and eight were elevated in *SD2* compared to *SD1*.

Table 4. Top 10 Random Forest metabolites distinguishing *SD1* vs. *DD* and *SD1* vs. *SD2*. **Bold** metabolites are elevated later stages.

Rank	*SD1/**DD***	*SD1/**SD2***
1	**Glutamine**	**Serine**
2	**Cholesterol**	**Choline Phosphate**
3	**Nicotinamide**	**Glutamate**
4	**Spermidine**	**Threonine**
5	**GPC**	**GPE**
6	**Malate**	Fumarate
7	**5'CMP**	**Glycerate**
8	**Glutamate**	**5-Oxoproline**
9	**N-Stearoyl-Sphingosine**	**Tryptophan**
10	**Serine**	Malate

4   Conclusions

Results of the earlier cytokine experiments indicated that *DD* immediately affected their concentrations; they remained stable (suggesting a more permanent impact) or returned to previous levels (suggesting that continued *DD* was required). We see examples of both cases. The evident rise in alpha-diversity in *DD* compared to *SD1* that was maintained throughout *SD2* indicates that the *DD* could be creating a long-term environment where more bacterial taxa can survive. Analysis of taxa abundance changes showed the taxa most immediately affected by the *DD* to be *Prevotella, Veillonella, Parvimonas, Rothia,* and *Corynebacterium.* The taxa *Prevotella, Veillonella,* and *Parvimonas* declined in *DD* and tended to stay low in *SD2,* while the taxa *Rothia* and *Corynebacterium* increased in *DD* and then seemed to return to previous levels in *SD2.* The taxa *Prevotella* and *Veillonella* are interesting as they make up core microbiota observed in *bronchiectasis,* a feature of A1AD [34]. This suggests that the *DD* can help to produce a long-term reduction in these populations. Additionally, the species *Corynebacterium pseudodiphtheriticum* has recently been touted as a possible lung probiotic as it differentially improves the immune response in mice to various lung infections [22]. This same study also indicated that a *colonization* of this species is necessary to achieve this improvement, which becomes interesting when we look at our networks. After *SD1, Corynebacterium* was completely disconnected and appeared to have no supporting taxa, but by the end of *DD* had formed a significant number of positive connections, that continued to grow throughout *SD2* even when its relative abundance dropped. This could indicate a progressive colonization of *Corynebacterium* throughout the augmentation therapy. Additionally the rise in this abundance in *DD,* with the corresponding return to *SD1* levels in *SD2,* could indicate the *DD* may be playing an important role in restoring balance after dysbiosis. It should not be ignored, however, that the heavy dynamics in the *Propionibacterium, Streptococcus, Actinomyces* and *Gammaproteobacterium* populations when going from *DD* to *SD2* that were not present when comparing *SD1* to *DD,* appear to indicate that going from a stronger dosage to a weaker had more effects on the overall microbial composition compared to going from a weaker to a stronger. *SD2* analysis indicates *Anaerococcus* may have an increasingly significant role in these dynamics.

Our volcano plots found the fewest important features when distinguishing *SD1* and *SD2* compared to *SD1/DD* and *DD/SD2,* which also supports the previous analysis of "returning to original levels". Viewing some of the important metabolites discovered by the volcano plots and PLS-DA, it does seem that the *DD* had some longer lasting effects, including the elevated 5'CMP (higher in *DD* and *SD2* compared to *SD1*) and the reduced 2-Aminoheptanoate (higher in *SD1* compared to *DD* and *SD2*). The long term rise in AMP in *SD2* compared to *SD1* could also be connected to the rise in the *Actinomyces* population, as could be the rise in Glutamate shown in Random Forest. 68% of *Actinomyces* reaction pathways in the PathwayTools [23] database involved at least one of these two metabolites. Glutamate is particularly interesting because of its importance for producing Glutamine [30], whose deficiency has heavily documented connections to several respiratory ailments, including the A1AD post-cursor COPD [28]. Interestingly, Glutamine was discovered as #1 when distinguishing *SD1* vs. *DD* (and it was elevated in *DD*), but only Glutamate (not Glutamine) was discovered as important in distinguishing *SD1* vs. *SD2.* This indicates that although a SD may be enough to increase Glutamate levels, the DD may be necessary to maintain high levels of Glutamine. We also note Random Forest to have discovered an interesting time series of important features involving Nicotinamide in *SD1-DD* (elevated in *DD*) followed by Malate in *SD1-SD2* (elevated in *SD1*), that parallels what we observe in the Krebs cycle [25] with Nicotinamide Adenosine Dinucleotide (NAD) molecules facilitating the conversion of Malate to CO_2 and Pyruvate. The increased presence of Nicotinamide in *DD*, followed by the decreased presence of Malate in *SD2*, could indicate a long-term increased Krebs cycle efficiency throughout augmentation therapy.

Our results provide a cursory overview of the observed effects of multiple steps of augmented therapy on the microbiome and metabolome of a small set of A1AD patients. Ultimately the microbiome and metabolome are interdependent systems, and their integrated dynamics involve a complex web of underlying interactions involving members from both systems. Therefore to establish adequate depth of knowledge to understand the causes of observable behaviors such as those we have presented, finding points of connection between their results will be fundamental (i.e. *Actinomyces* with AMP and Glutamate). Connections could also be established with some of the specific proteases found to fluctuate in concentration in the augmentation therapy [5], including serine proteases, neutrophil-derived elastase, cathepsin G, and Aα-Val. This multi-omics analysis will necessitate further data collection across multiple metabolomics databases (i.e. PathwayTools, KEGG [21]) to continue to complete this puzzle. It also may demand moving from 16 S sequencing to shotgun metagenomics, to reveal a more complete genetic profile of taxa within each sample [43]. Additionally, similarity metrics for signed and weighted networks such as SASCOS++ [12] would provide a more quantitative metric for comparing networks, as opposed to our more qualitative approaches. We also recommend centrality [11] analysis to view important members of both systems; and potentially heterogeneous networks to view relationships across systems. Signed Bayesian Networks [33] could prove very useful in this area, due to their implications for causality. We see the analysis presented here as a starting point for more in-depth biological interpretation, wet lab experiments, and causal analysis.

Acknowledgment. GN and KM were supported by National Institute of Health (580 #1R15AI128714-01). GN was also supported by Department of Defense (581 #W911NF-16-1-0494) and the National Institute of Justice (582 #2017-NE-BX-0001). GN, KM and MC were supported by the Florida Department of Health (FDOH 09KW-10) and the Alpha-One Foundation. TC received support from NVIDIA and Florida International University. The authors also thank colleagues from the BioRG for many useful discussions.

References

1. Barker, M., Rayens, W.: Partial least squares for discrimination. J. Chemometr. **17**(3), 166–173 (2003)
2. Bolger, A.M., Lohse, M., Usadel, B.: Trimmomatic: a flexible trimmer for Illumina sequence data. Bioinformatics **30**(15), 2114–2120 (2014)
3. Brebner, J.A., Stockley, R.A.: Recent advances in -1-antitrypsin deficiency-related lung disease. Expert Rev. Respir. Med. **7**(3), 213–230 (2013)

4. Breiman, L.: Random forests. Mach. Learn. **45**(1), 5–32 (2001). https://doi.org/10.1023/A:1010933404324
5. Campos, M.A., et al.: The biological effects of double-dose AAT augmentation therapy. Am. J. Respir. Crit. Care Med. **200**(3), 318–326 (2019)
6. Campos, M.A., et al.: Safety and pharmacokinetics of 120 mg/kg versus 60 mg/kg weekly intravenous infusions of alpha-1 proteinase inhibitor in A1AD: a multicenter, randomized, double-blind, crossover study (SPARK). COPD: J. Chronic Obstr. Pulm. Dis. **10**(6), 687–695 (2013)
7. Caporaso, J.G., et al.: QIIME allows analysis of high-throughput community sequencing data. Nat. Methods **7**(5), 335–336 (2010)
8. Chao, A., Chazon, R.L., Colwell, R.K., Shen, T.: A new statistical approach for assessing similarity of species composition with incidence and abundance data. Ecol. Lett. **8**, 148–159 (2005)
9. Chmiel, J.F., et al.: Antibiotic management of lung infections in cystic fibrosis. I. The microbiome, methicillin-resistant staphylococcus aureus, gram-negative bacteria, and multiple infections. Ann. Am. Thorac. Soc. **11**(7), 1120–1129 (2014)
10. Chung, K.F.: Potential role of the lung microbiome in shaping asthma phenotypes. Ann. Am. Thorac. Soc. **14**(S5), S326–S331 (2017)
11. Cickovski, T., Peake, E., Aguiar-Pulido, V., Narasimhan, G.: ATria: a novel centrality algorithm applied to biological networks. BMC Bioinformatics **18**, 239–248 (2017). https://doi.org/10.1186/s12859-017-1659-z
12. Derr, T., Wang, S., Wang, C., Tang, J.: Relevance measurements in online signed social networks. In: The 14th International Workshop on Mining and Learning with Graphs, London, UK, August 2018
13. Edgar, R.C.: USEARCH and UCLUST algorithms. Bioinformatics **26**(19), 2460–2461 (2010)
14. Fernandez, M., Riveros, J.D., Campos, M., Mathee, K., Narasimhan, G.: Microbial "social networks". BMC Genomics **16**(11), 1 (2015)
15. Frey, B.J., Dueck, D.: Clustering by passing messages between data points. Science **315**(5814), 972–976 (2007)
16. Fruchterman, T.M.J., Reingold, E.M.: Graph drawing by force-directed placement. Softw. Pract. Exper. **21**(11), 1129–1164 (1991)
17. Heid, C.A., Stevens, J., Livak, K.J., Williams, P.M.: Real-time quantitative PCR. Genome Res. **6**, 986–994 (1996)
18. Hubbard, R.C., Sellers, S., Czerski, D., Stephens, L., Crystal, R.G.: Biochemical efficacy and safety of monthly augmentation therapy for A1AD. JAMA **260**(9), 1259–1264 (1988)
19. Hunt, J.M., Tuder, R.: Alpha-1 anti-trypsin: one protein, many functions. Curr. Mol. Med. **12**(7), 827–835 (2012)
20. John, J.S.: SeqPrep (2019). https://github.com/jstjohn/SeqPrep
21. Kanehisa, M., Goto, S.: KEGG: kyoto encyclopedia of genes and genomes. Nucleic Acids Res. **28**(1), 27–30 (2000)
22. Kanmani, P., et al.: Respiratory commensal bacteria Corynebacterium pseudodiptheriticum improves resistance of infant mice to respiratory syncytial virus and Streptococcus pneumonias superinfection. Front. Microbiol. **8**, 1613 (2017)
23. Karp, P., Paley, S., Romero, P.: The pathway tools software. Bioinformatics **18**(suppl 1), S225–S232 (2002)
24. Kozich, J.J., Westcott, S.L., Baxter, N.T., Highlander, S.K., Schloss, P.D.: Development of a dual-index sequencing strategy and curation pipeline for analyzing amplicon sequence data on the MiSeq illumina sequencing platform. Appl. Environ. Microbiol. **79**(17), 5112–5120 (2013)

25. Krebs, H.A.: The citric acid cycle. Biochem. J. **34**(3), 460–463 (1940)
26. Laurell, C.B., Eriksson, S.: The electrophoretic Alpha1-Globulin pattern of serum in Alpha1 antitrypsin-deficiency. Scand. J. Clin. Lab. Invest. **15**(12), 132–140 (1963)
27. O'Dwyer, D.N., Dickson, R.P., Moore, B.B.: The lung microbiome, immunity and the pathogenesis of chronic lung disease. J. Immunol. **196**(12), 4839–4847 (2017)
28. Oliveira, G.P., de Abreu, M.G., Pelosi, P., Rocco, P.R.M.: Exogenous glutamine in respiratory diseases: myth or reality? Nutrients **8**(2), 76 (2016)
29. Padmanaban, A., Inche, A., Gassman, M.: High-throughput DNA sample QC using the agilent 2200 tapestation system. J. Biomol. Tech. **24**(Suppl), S41 (2013)
30. Purves, D., et al.: Neuroscience, 4th edn. Sinauer Associates, Sunderland (2008)
31. Quast, C., et al.: The SILVA ribosomal RNA gene database project: improved data processing and web-based tools. Nucleic Acids Res. **41**, D590–D596 (2013)
32. Ruiz-Perez, D., Guan, H., Madhivanan, P., Mathee, K., Narasimhan, G.: So you think you can PLS-DA? In: International Conference on Computational Advances in Bio and Medical Sciences, ICCABS 2018. IEEE (2018)
33. Sazal, M., Ruiz-Perez, D., Cickovski, T., Narasimhan, G.: Inferring relationships in microbiomes from signed bayesian networks. In: International Conference on Computational Advances in Bio and Medical Sciences, ICCABS 2018. IEEE (2018)
34. Schafer, J., Griese, M., Chandrasekaran, R., Chotirmall, S.H., Hartl, D.: Pathogenesis, imaging and clinical characteristics of CF and non-CF bronchiectasis. BMC Pulm. Med. **18**, 79 (2018). https://doi.org/10.1186/s12890-018-0630-8
35. Segata, N., et al.: Metagenomic biomarker discovery and explanation. Genome Biol. **12**(6), R60 (2011)
36. Sender, R., Fuchs, S., Milo, R.: Revised estimates for the number of human and bacteria cells in the body. PLoS Biol. **14**(8), e1002533 (2016)
37. Shannon, C.E.: A mathematical theory of communication (1963). Bell Syst. Tech. J. **27**, 379–423 (1948)
38. Sharp, H.L., Bridges, R.A., Krivit, W., Freier, E.F.: Cirrhosis associated with A1AD: a previously unrecognized inherited disorder. Trans. Res. **73**(6), 934–969 (1969)
39. Spearman, C.: The proof and measurement of association between two things. Am. J. Psychol. **15**(1), 72–101 (1904)
40. Steemers, F.J., Gunderson, K.L.: Illumina, Inc. Pharmacogen **6**(7), 777–782 (2005)
41. Stoller, J.K., Aboussouan, L.S.: 1-antitrypsin deficiency · 5: intravenous augmentation therapy: current understanding. Thorax **59**(8), 708–712 (2004)
42. Surette, M.G.: The cystic fibrosis lung microbiome. Ann. Am. Thorac. Soc. **11**(S1), S61–S65 (2014)
43. Tyson, G.W., et al.: Community structure and metabolism through reconstruction of microbial genomes from the environment. Nature **428**, 37–43 (2004)
44. Xia, J., Sinelnikov, I.V., Han, B., Wishart, D.S.: MetaboAnalyst 3.0—making metabolomics more meaningful. Nucleic Acids Res. **43**(W1), W251–W257 (2015)

A Multi-hypothesis Learning Algorithm for Human and Mouse miRNA Target Prediction

Mohammad Mohebbi[1]([envelope])[iD], Liang Ding[2], Russell L. Malmberg[3][iD], and Liming Cai[4]

[1] Department of Computer Science, Appalachian State University, Boone, NC 28608, USA
`mohebbim@appstate.edu`
[2] St. Jude Children's Research Hospital, Memphis, TN 38105, USA
`Liang.Ding@stjude.org`
[3] Department of Plant Biology, The University of Georgia, Athens, GA 30602, USA
`malmberg@uga.edu`
[4] Department of Computer Science, The University of Georgia, Athens, GA 30602, USA
`cai@cs.uga.edu`

Abstract. MicroRNAs (miRNAs) are small non-coding RNAs that play a key role in regulating gene expression and thus in many cellular activities. Dysfunction of cells in these tasks is correlated with the development of several kinds of cancer. As the functionality of miRNAs depends on the location of their binding on their targets, binding site prediction has received a lot of attention in the last several years. Despite its importance, the mechanisms of miRNA targeting are still unknown. In this paper, we introduce an algorithm that partitions miRNA target duplexes according to hypotheses that each represents a different mechanism of targeting. The algorithm, called multi-hypothesis learner, examines all possible hypotheses to find out the optimum data partitions according to the performance of these hypotheses for miRNA target prediction. These hypotheses were then utilized to build a superior target predictor for miRNAs. Our method exploited biologically meaningful features for recognizing targets, which enables establishment of hypotheses that can be correlated with target recognition mechanisms. Test results show that the algorithm can provide comparable performance to state-of-the-art machine learning tools such as RandomForest in predicting miRNA binding sites. Moreover, feature selection on the partitions in our method confirms that the partitioning mechanism is closely related to biological mechanisms of miRNA targeting. The resulting data partitions can potentially be used for *in vivo* experiments to aid in discovery of the targeting mechanisms.

This work was supported in part by NIH grant (award No: R01GM117596), as a part of Joint DMS/NIGMS Initiative to Support Research at the Interface of the Biological and Mathematical Sciences, and NSF IIS grant (award No: 0916250).

I. Măndoiu et al. (Eds.): ICCABS 2019, LNBI 12029, pp. 102–120, 2020.
https://doi.org/10.1007/978-3-030-46165-2_9

Keywords: miRNA · Machine Learning · miRNA target prediction · Multi-Hypothesis Learning · Data partitioning

1   Introduction

MicroRNAs (miRNAs) are short RNA sequences of approximately 22 nucleotides that inhibit or repress gene expression. They perform as a guide to bind the RISC (RNA Induced Silencing Complex) to sequence specific locations on mRNAs to silence them [1]. These specific locations are called target sites and discovering the functionality of each miRNA depends on recognition of its target sites. MiR-NAs can control many critical cell processes such as proliferation, differentiation, cell death, growth control and developmental timing [22]. Dysfunction of miR-NAs could lead to tumor development and cancer in organs such as lung, brain, colon and breast in addition to causing hematopoietic cancers [16].

Despite the importance of miRNAs the detailed mechanism of miRNA target binding is poorly known. Lab experiments for finding targets are very slow and costly, therefore there is a huge demand for computational approaches. In the last decade dozens of algorithms, with a variety of approaches and techniques, have been developed. These methods are either specific for a few species or general for any kind. Methods for vertebrates include TargetScan and TargetScanS [20,21], miRanda [9,17], DIANA-microT [19] and for flies RNAhybrid [30]. Some general tools are miTarget [18] and MicroInspector [31].

The early computational approaches for target recognition were rule based, i.e., they had a set of discriminative rules derived from experimental and biological knowledge, such as MFE (Minimum Free Energy), duplex binding pattern, or target accessibility. Some popular rule based tools are RNAhybrid, TargetScan, miRanda and MirBooking [37]. MirBooking is one of the recent rule based methods that simulates the miRNA and mRNA hybridization competition and cellular conditions to improve the accuracy of target prediction. In the last several years, with the increase of relevant data sets, data driven methods have been attempted. These methods use sophisticated machine learning and statistical models to learn more discriminative features for target identification [39]. Some popular data driven tools are TargetSpy [33], miRanda-mirSVR [3] and Avishkar [12]. However, such methods have yet to resolve the issue of high false positive rate. The innovation of more advanced sequencing techniques, and therefore more precise data sets, along with recent advances in machine learning methods, could lead to the development of more accurate algorithms.

The miRNA targeting process has not been well understood; biologists are especially interested in approaches that may provide insights about the mechanisms of target recognition. Recent experimental studies of miRNA targeting reveal that there are multiple and different mechanisms for this process, while the earlier belief was merely based on seed match of miRNA and target site sequences [6]. Currently, it is still not clear how many different and exclusive mechanisms guide miRNA targeting, therefore computational models which not only work well but also give insight into the biological mechanisms, are very desirable. Some machine learning techniques such as Bagging and Boosting or Random Forest aim to learn multiple hypotheses from the input data, but they do not provide any clue to check if these hypotheses are biologically meaningful or not. Biologically meaningful features here means those characteristics that have been experimentally confirmed to be part of a miRNA targeting mechanism, such as appearance of Adenine at the far 3' end of a target site.

In this work, we introduce a multi-hypothesis learning (MHL) method that builds specific models and hypotheses for each mechanism of miRNA targeting. We exploit the models for two main purposes; first to build an miRNA target prediction algorithm with a superior performance, and second to partition the miRNA target dataset in a biologically meaningful way that could be used for further understanding the targeting mechanisms or to discover new target determinants. To verify our approach we evaluated our method on human and mouse data. The results show that the partitioning is indeed biologically meaningful. Moreover, significant performance improvements on target prediction confirms learning multiple hypotheses can help outperform top machine learning algorithms such as RandomForest. Feature analysis of the partitions produced by MHL reveals interactions in miRNA and target duplex that are verified by the biology literature. This supports our conjecture that MHL could aid to mine meaningful features, which could be used as part of *in vivo* experiments.

2   Data Sets

The success of data driven methods critically relies on the quality of the data. To build the most accurate models and the most realistic evaluations, we extracted our data from mirTarBase [15], one of the most up-to-date data sets and the most referenced resource for miRNA target prediction research. In particular, mirTarBase contains more than 360,000 experimentally validated miRNA-target duplexes from 18 different species. We are mostly interested in testing our machine learning method with both human and mouse records.

From mirTarBase, human and mouse miRNA-target duplexes were extracted whose secondary structures have been provided in research articles. Such duplexes were selected as positive samples for our method. However, negative samples are not directly available. Theoretically, any stretch of an appropriate length other than the real target in the 3'-UTR of a targeted mRNA gene can be considered a negative target of the corresponding miRNA. We randomly selected ten locations in the 3'-UTR of a targeted mRNA gene to pick up the negative samples for each positive sample with a ratio of ten to one. Each sample is a pair of miRNA sequence of length 22 and a site sequence of length 25 which is real target site for positive samples or a negative site that is not a target for the miRNA.

2.1   Test Set and Training Set

We have one training set and two test sets. The human dataset is split 80% to 20% into the training set and human test set. All mouse data composes our second test set. In the human data extracted from mirTarBase, there are 322 unique miRNAs, 3651 target site sequences and 3722 pairs of miRNA and target sites. On average, each miRNA has >10 targets sites. If we randomly select test set samples from the whole database, the odds of having many miRNAs in both test and training sets is high. To avoid such overlaps and to have the most reliable test set, we indexed pairs of miRNA and target sites by miRNA sequence. In addition, to make a test set with a similar distribution to that of the whole dataset, we sort samples by miRNA sequences, put four consecutive (based on the sorting order) miRNA sequences and all their target and non-target sites in the training set, then one miRNA sequence and all its targets and non-targets into the human test set and so on. In this way and in terms of miRNA sequence, not only do the human test set and training set have no overlaps but also the test set has very similar distribution to that of the whole database. Both test sets and training sets have ratio of 1:10 for positive vs. negative. The human test set consists of 6127 samples (557 positives vs. 5570 negatives), and the total size of the mouse test set is 517.

3   The Model and Method

In this section, we introduce a feature selection approach which not only is more efficient for miRNA targeting than previous data mining feature selection methods, but also it is biologically meaningful too. Data mining algorithms could not be applied directly on this problem because each sample is composed of sequences of miRNA and target, and the miRNA sequence is identical among its positive(s) and its negative samples. Hence when we ran Weka [38], a data mining package, to extract features, all miRNA sequence nucleotides were excluded from selected features set. To cope with this problem, features must be defined based on correlations of miRNA and its target nucleotides rather than merely on sequences of nucleotides. In addition, and to incorporate biological knowledge of miRNA targeting, we extracted features from the secondary structure of a duplex associated with every sample.

We customized RNAfold [24], a widely used secondary structure prediction tool for RNA sequences, to predict a specific structure for a pair of miRNA and target sequences given the most recent discoveries on the mechanisms of miRNA targeting. Biologically, sequences of miRNA and target sites should not make base-pairs with themselves but with the other sequence. In general, RNAfold could predict structures in which miRNA or target site sequences might bind to themselves. To avoid this problem, and include information about *in vivo* process of miRNA target binding, we tuned RNAfold to predict the structure of duplex based on rules we collected from biological literatures explaining the actual mechanisms of miRNA targeting.

The seed of an miRNA consists of the nucleotides number 2 to 8 from the 5' end of the miRNA [21]. It is believed that the process of nucleotide binding between the miRNA and its mRNA target starts from this region [32]. When the binding in the seed region is continuous for 6 to 8 bps, it is called a canonical seed;

otherwise it is called non-canonical [23]. Though the seed binding is considered the most important identifier for miRNA targets in mammals [29], a recent study shows it is not the only mechanism for miRNA targeting [6]. To have a more comprehensive model, we considered correlations that occur not just in the seed region but also in all other regions across a miRNA and its target.

3.1   RNAfold Customization and Feature Selection

RNAfold for a given biomolecule sequence of nucleotides predicts the most stable secondary structure of the molecule. To use it for predicting miRNA and target site duplexes, we concatenated miRNA and target site sequences with a subsequence of length four 'X's in between, as shown in Fig. 1. This sequence of length four is the shortest sequence that we could add and still get the same MFE (minimum free energy) for the structure as the MFE we get from RNAcofold [24] when it predicts the duplex between miRNA and target site.

RNAfold can have a *constraints* file as an input parameter, to enforce the structure prediction process to occur based on a user's domain knowledge. Here we set these *constraints* for the miRNA targeting mechanism, to include rules for base-pairs which are biologically expected to happen in seed, and rules prohibiting miRNA nucleotides from binding to miRNA itself. Similarly, there are rules avoiding target site sequence to bind over itself. We applied all these rules for duplexes with canonical seeds, while releasing seed base pairing constraints for non-canonical seeds.

Biological experiments and *in vivo* methods reveals several mechanisms for miRNA targeting [6]. The earliest discovered and the most dominant method of

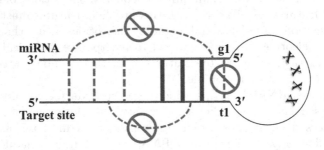

Fig. 1. RNAfold customization for a miRNA target duplex; sequences of miRNA and target site are concatenated with a subsequence of length four 'X's in between. Base pairing among nucleotides of one sequence is prohibited, either for miRNA or the target site. A structural study on the mechanism of miRNA targeting [32] reveals that the nucleotide in t1 goes into a pocket inside the Argonaute protein structure and does not pair with the corresponding nucleotide on miRNA. we added a constraint to prevent t1 from such a base pairing with the miRNA. Base-pairs (in purple color) are enforced through the customization if the corresponding nucleotides are complimentary matching. black dashed lines show possible locations of valid base-pairs, and we let RNAfold to predict them. (Color figure online)

targeting was based on seed matching [10, 20]. In this mechanism, miRNA carried by the Argonaute protein makes initial base pairs in the seed area. These bindings open the groove of Argonaut molecule to accommodate the target site [32]. To customize RNAfold for predicting duplexes in a similar fashion, we aligned the seed part of miRNA with nucleotides 2 to 8 from 3' side of target and pair these bases that can match to each other mutually; i.e. Adenine (A) to Uracil (U), Cytosine (C) to Guanine (G), Guanine to Uracil and vice versa.

The secondary structure predicted by RNAfold is a list of base pairs between nucleotides in an miRNA and its target site. To apply machine learning based algorithms on the structure, we needed to map it to a vector of numbers. These base pairing features are nominal and to convert them into numerical values while maintaining their independence, we encoded them with the One-hot-encoding (OHE) approach [34]. Biologically there is no significance to the ordering among six different base pairs; to keep this independence, we encoded each matching base pair with one bit, totaling six bits, and one extra bit for mismatches or no base pairs. One and only one of these seven bits is *hot* or one at a time. In an miRNA duplex structure there might be bulges on either miRNA or target sequence. To incorporate this information into the vector we added two integer values indicating size of bulges on the miRNA and on the target site, adjacent to each nucleotide. These values are zero if there is no bulge in the structure next to the current nucleotide. In total there are 9 features per each miRNA nucleotide.

Experimental studies on human miRNA targets showed Adenine is a very frequent base at the far 3' end of a target site, i.e. at t1 [20, 32]. To add this biological preference to features set, we added four bits corresponding to A, C, G and U at t1. A study on the structural basis of miRNA targeting [32], revealed that the nucleotide in t1 goes into a pocket inside the Argonaute protein structure and does not pair to the corresponding nucleotide on miRNA, i.e. g1, which is the first nucleotide on 5' end of the miRNA. Therefore, to reduce the size of features set, we excluded g1 from being encoded. A factor indicating stability of a structural binding is MFE, we included it as the last feature. We fixed the length of miRNAs to 22 nucleotides, but g1 is not considered, therefore the total number of features for each sample is 194 or $(1 + 4 + 21 * 9)$. If the length of miRNA is larger than 22, the sequence is trimmed to 22 from 3' side of miRNA. This procedure is illustrated inside dashed area of Fig. 2. Our MHL algorithm, to be introduced in the next section, treats each sample as a vector of these 194 features and learns several hypotheses each corresponding to a different miRNA targeting mechanism. Figure 2 shows all components of our bundle algorithm including the feature selection part and the MHL algorithm.

3.2 The Algorithm

The idea of the algorithm is to divide the dataset into two disjoint subsets sb_1 and sb_2 such that these two subsets have similar distributions of labels or classes. It learns the major pattern in sb_1 with classifier c_1 and stores it as model m_1. Then

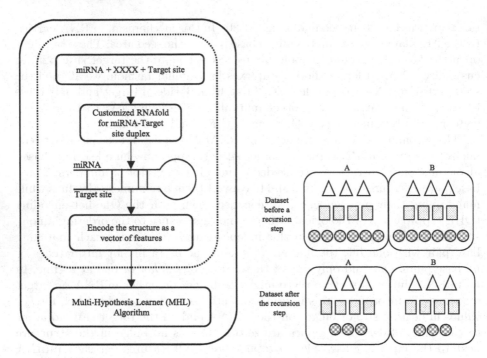

Fig. 2. Our bundle algorithm; it gets two sequences of miRNA and target, concatenates them with subsequence 'XXXX'. The resulting sequence is passed to RNAfold that we customized as explained in Fig. 1. The customized RNAfold predicts the secondary structure of the miRNA and the target duplex. The structure is encoded as a vector of features and passed down to MHL (Shown in Fig. 3).

Fig. 3. The illustration of the MHL recursion algorithm; Dataset D is split to subsets A and B, then a classifier c_i is trained on A. It captures the dominant pattern, here *circles*. The trained model can detect the similar pattern in B, i.e. *circles*; these form the *can_decide* set and are removed from B. The remaining data in B, i.e. the *cannot_decide* set is combined with A and split again. In each recursion a model, or a hypothesis, is learned for the current dominant pattern of data. The recursive process continues until no dominant pattern is left, then the last model is trained on the remaining samples and the process stops.

it partitions sb_2 based on m_1's performance into two parts *can-decide* or *cannot-decide* samples. The subpartition *can-decide* contains instances where m_1 can predict their labels with confidence while the other subpartition includes those that m_1 is not sure about their classification. Subsets *cannot_decide* and sb_1 are merged to yield a new training set. The process is repeated recursively on the new merged set until no further partitioning into *can-decide* and *cannot_decide* is possible.

The algorithm consists of two main parts: Trainer and Tester. Trainer gets the training set T_0, a classifier set C, and a desired sensitivity and specificity: _sen and _spec. During a recursive procedure, Trainer builds regression models, i.e. hypotheses, specific for different patterns of data, which are observed in the input training set T_0. It also stores each produced model M along with two thresholds T_{up} and T_{down} for the Tester part. For every sample evaluated by M with a value $\geq T_{up}$, it would be classified as positive while labeled as negative if it's evaluation value is $< T_{down}$. The model M guarantees the desired sensitivity and specificity _sen and _spec for the *can-decide* partition. When the evaluation value is between T_{down} and T_{up}, the model does not classify the sample and it would be added to the *cannot_decide* set.

3.3 Trainer

Trainer consists of three functions: $Splitter()$, $Model\_Builder()$, and $Threshold$ $\_Finder()$. The $Splitter$ (D, C) gets a dataset D and a set of classifiers C as input. Classifiers are Weka training modules accessible through its API (Application Programmable Interface). The $Splitter()$ function splits the input set D into two subsets A and B by the Stratification method [35] to maintain the same ratio of positive samples versus negatives in these subsets as it is in D. A and B are disjointing and complement of each other corresponding to D, i.e. $A \cup B = D$. By calling the function $Model\_Builder(c_i, A, B)$, the model m_i is built by classifier c_i on dataset A. In addition, the function splits B into *can-decide* and *cannot_decide* subsets by evaluating m_i on B samples. A is merged with *cannot_decide* and is returned as D_{new1}, the new training set. Then the process is recursively repeated on this new set. To avoid any bias toward the way we split the data by the Stratification method: we swap the position of A and B then repeat the process.

Depending on how high the thresholds _sen and _spec are chosen, the $Model$ $\_Builder()$ function may not be able to build such a model and might not return a new training set. In such a case, it returns the same set as the input training set, indicating it failed to build the desired model. Given this condition, function $Splitter()$ builds a model with c_i on the input training set D and stops.

There are two thresholds associated with each trained model; T_{up} and T_{down}. The algorithm $Threshold$ $\_Finder$ computes these thresholds such that the model m_i had a given and desired sensitivity and specificity _sen and _spec. The higher sensitivity and specificity resulted in larger *cannot_decide* subset in B. We denote the *cannot_decide* subset as β.

Algorithm 1: Splitter (D, C)

---
1 foreach *classifier* $c_i \in C$ do
2 | Split D into two subsets A and B by the Stratification method;
3 | D_{new1} = Model_Builder(c_i, A, B) ;   /* Model m_i is stored as m_{i_a}. */
4 | if $|D_{new1}| < |D|$ then
5 | | Splitter (D_{new1}, C);
6 | end
7 | D_{new2} = Model_Builder($c_i,$ B, A) ;   /* Model m_i is stored as m_{i_b}. */
8 | if $|D_{new2}| < |D|$ then
9 | | Splitter (D_{new2}, C);
10 | end
11 | if $|D_{new1}| == |D|$ **OR** $|D_{new2}| == |D|$ then
12 | | train c_i with D, store it as m_i and stop;
13 | end
14 end

Algorithm 2: Model_Builder(c_i, s_a, s_b)

---
1 Train *classifier* c_i on set s_a, store the trained model as m_{i_a};
2 Evaluate set s_b by model m_{i_a};
3 Store the evaluations as a list L_b of $Pair(sample.label, sample.evaluation)$;
4 Pair (T_{down}, T_{up}) = Threshold_Finder($L_b, \_sen, \_spec$) ;   /* find T's satisfying sensitivity and specificity. */
5 β = subset of s_b that evaluated as $>= T_{down}$ and $< T_{up}$;   /* the *cannot_decide* subset. */
6 Store the model m_{i_a} with (T_{down}, T_{up});
7 Store $(s_b - \beta)$ as an ARFF file ;   /* $s_b - \beta$ is the *can_decide* subset, store it for further feature analysis. */
8 Return $s_a \cup \beta$.

Algorithm 3: Threshold_Finder $(L_b, \_sen, \_spec)$

---
1 $T_{up} = 1, T_{down} = 0$;
2 Votes [] $= \emptyset$;
3 do
4 | foreach *pair* $p_i \in L_b$ do
5 | | if $p_i.evaluation >= T_{up}$ then
6 | | | Votes[p_i] = *positive*;
7 | | end
8 | | if $p_i.evaluation < T_{down}$ then
9 | | | Votes[p_i] = *negative*;
10 | | end
11 | end
12 | Compute $\_sen_{tmp}$ and $\_spec_{tmp}$ for Votes[];
13 | if $\_sen_{tmp} < \_sen$ **OR** $\_spec_{tmp} < \_spec$ then
14 | | stop and break;
15 | end
16 | $T_{down} += \Delta$;   /* $\Delta = 0.05$ */
17 | $T_{up} -= \Delta$;
18 while $T_{down} < T_{up}$;
19 Return $Pair(T_{down}, T_{up})$;

3.4 Tester

The Tester procedure loads all model files, m_i's, from the training step into the memory and when a new and unlabeled sample is given for evaluation, all models examine the sample. If a model evaluates the sample with a value between T_{down} and T_{up} then it does not vote, otherwise it votes with confidence as positive if the value is $\geq T_{up}$ and as negative for the value $<T_{down}$. Each vote associated with a weight, which is the size of the dataset used to build the model. The weighted average of all votes is returned as the final prediction. If all classifiers' evaluation values are between T_{down} and T_{up}, that means there is no vote for the sample and it is predicted with label zero.

4 Results and Discussion

In the training set, there might be several patterns of miRNA targeting, here we denote them symbolically by *circles*, *squares*, *triangles*, and etc. as an example shown in Fig. 3. Initially, *circles* are the dominant pattern. The MHL algorithm divides it to subsets A and B. Classifier c learns *circles* pattern when it runs over subset A and creates a model for *circles*, i.e. m_c. Evaluating B with m_c divides B into two partitions; samples decidable by m_c, i.e. the *can_decide* set, here *circles*, and samples that m_c is not sure about, called *cannot_decide* set. *Circles* are removed from B because we have m_c that can detect them, but the rest of B are not recognizable by m_c so they are added to A to form a new training set. Now, in the new training set *squares* are the dominant pattern and in next recursion step, a model is built for them. This recursion will continue until all patterns are learned or there is no dominant pattern left. In later case, a model for the remaining samples is created by c and recursion stops.

4.1 Test of Our Multi-hypothesis Learning (MHL) Bundle Algorithm

To test the effectiveness of our algorithm, we compare the Area Under the Curve (AUC) of different Machine Learning (ML) models from the Weka package [38] versus our Multi-Hypothesis Learning (MHL) bundle algorithm. Table 1 and Table 2 present these comparison results on the human (HSA) and mouse (MMU) test sets, respectively. Columns of these tables are classifier(s) name, parameter sensitivity and specificity i.e. _sen and _spec for MHL, AUC of ML models, and AUC of our algorithm when the same ML classifier used as underlying model in the MHL.

These tables show that our algorithm is effective, especially when the ML model does not perform well on a test set. The tables show that our method MHL has the best performance with Linear Regression. For the human test data set, the AUC of this model for sequences of miRNA and target sites is 0.69 and, when the samples are evaluated by MHL, it increases the AUC to 0.93. The algorithm improves the AUC by 0.24 which is the highest increase over the ML

Table 1. Area Under the Curve (AUC) of different Machine Learning (ML) models versus our Multi-Hypothesis Learning (MHL) algorithm on HSA (Human) test set, $|HSA| = 6129$ samples. The last two rows of the table show performance of miRanda [4] and RNAybrid [30] on the test set [28]. RNAhybrid and miRanda performed similarly and MHL surpasses these methods due to two main reasons; first RNAhybrid and miRanda are rule-based while MHL is data-driven, and second MHL hypotheses are trained on experimentally verified samples from mirTarBase records. miRNA duplexes with $non-canonical$ seeds [23] might have higher MFE than duplexes with $canonical$ seeds and MFE based target prediction tools such as RNAhybrid and miRanda may not be able to detect them. On the other side, MHL learned specific hypotheses for $non-canonical$ samples, and this gives a unique advantage to MHL algorithm to outperform these rule-based methods.

Classifier(s)	ML models	_sen/_spec	MHL algorithm	Improvement
Random Tree	0.59	85/85	0.78	0.19
DecisionStump	0.58	85/85	0.75	0.17
REPTree	0.82	85/85	0.93	0.11
RandomForest	0.92	85/85	0.92	0.00
ANN	0.66	90/90	0.83	0.17
LinearRegression	0.69	85/85	0.93	0.24
M5P	0.85	90/90	0.93	0.08
RandomTree& DecisionStump	N/A	85/85	0.84	N/A
REPTree& RandomForest	N/A	80/80	0.93	N/A
miRanda	0.50			
RNAhybrid	0.50			

model itself. It seems the highest achievable AUC for our training and human test sets is 0.93; Random Forest is the only classifier that our algorithm does not have improvement over it as it already performed very well with AUC of 0.92. The effectiveness of the MHL algorithm is more obvious when it enables mediocre-performed classifiers such as REPTree, Linear Regression and M5P to beat Random Forest. Classifiers performing poorly on this test set such as Random Tree, Decision Stump and Artificial Neural Networks (ANN) can also be used in the MHL module to deliver a performance of 0.75 to 0.83 in AUC. The AUC of ML models has an average of 0.73 with a standard deviation 0.13 while our algorithm can perform with AUC average of 0.87 and standard deviation 0.08. Our algorithm improves the average performance by 14%. To compute the averages, we did not consider combined classifiers, i.e., the last two rows in both tables are left blank.

We conjectured that some patterns of data could be learned better with a classifier than with the others, therefore we recursively searched all possible combinations of partitioning with different classifiers. To test this idea, we used combinations of RandomTree and DecisionStump in MHL and surprisingly, the AUC increased to 0.84 from 0.78 for RandomTree and from 0.75 for Decision-Stump, when these two classifiers were used individually in MHL. This combination outperformed these models by 0.25 if they were used independently and without MHL.

Table 2. Area Under the Curve (AUC) of different Machine Learning (ML) models versus our Multi-Hypothesis Learning (MHL) algorithm when they were trained on Human and tested on MMU (Mouse) test set, $|MMU| = 517$ samples.

Classifier(s)	ML models	$\_sen/\_spec$	MHL algorithm	Improvement
RandomTree	0.74	85/85	0.95	0.21
DecisionStump	0.48	85/85	0.61	0.13
REPTree	0.93	85/85	1.00	0.07
RandomForest	0.98	85/85	0.99	0.01
ANN	0.75	90/90	0.97	0.22
LinearRegression	0.52	85/85	0.99	0.47
M5P	0.95	90/90	0.97	0.02
RandomTree& DecisionStump	N/A	85/85	0.94	N/A
REPTree& RandomForest	N/A	80/80	1.00	N/A

Human and mouse branched from a common ancestor about 80 million years ago. They have similar genomes and virtually the same set of genes [11]. Therefore, it is of interest to train a model by human genomic data and test it on mouse data sets. Similarly we ran the same model used for testing human data, on mouse dataset and the results are shown in Table 2. Our algorithm improves over all ML classifiers and the maximum improvement again is for Linear Regression with an increase of 0.47 in AUC. The average performance of our algorithm is 0.93 with a standard deviation of 0.14 while ML models have an average of 0.76 and standard deviation 0.14. The algorithm average performance surpasses over ML methods average by 0.17.

Contrasting Table 1 with Table 2 shows that the ML models and our algorithm performs slightly better on mouse than on human data. The similar performance of these models on both mouse and human test sets suggests that miRNA target duplexes and targeting mechanism features are evolutionary conserved across both species. Some miRNAs have conserved sequences among human and mouse, consequently there might be an small portion of samples with similar sequences in both the (human) training set and the mouse test set. This could be the reason for a larger performance improvement on the mouse versus the human test set. In the human test set, we reduced the chance of miRNAs with similar sequences in both training and test sets by sorting miRNAs and dividing them between training and test sets alternatively with the given ratio of 10 to 1. Moreover, the test set was subtracted from the training set to avoid any chance of overlaps between sets. In the mouse test set, any correlation with the human training set is due to evolutionary conservation. Therefore, a small portion of miRNAs which are common between mouse and human, may lead to sample similarity between human training set and mouse test set.

Some machine learning packages, for example RandomForest, or GraBmiTarget [28] a hybrid model of a graph and an SVM (Support Vector Machine), have AUC performance as high as 0.92, but the unique advantage of MHL over all other methods is to provide a clue into the available data and to partition the data into data sets that are biologically meaningful clusters.

To examine if the clusters provided by MHL may have a different biological meaning than the set of all training data, we compare features selected by CFS (Correlation based Feature Selection) subset feature selection [14] from all training data versus the features extracted with the same method from the clusters. From the Weka package we ran the CFS method on all training data; the features extracted are shown in the first column of Table 3. The features are in the format of i_BP, $i\_MisMatch$, $i\_Bulge\_on\_miRNA$ or $i\_Bulge\_on\_target$, where i is the nucleotide index starting from 1 on 5' side of miRNA. BP composed of two letters X and Y representing a canonical base pair between nucleotides X and Y. As an example 2_AU depicts a base pair $A - U$ where A is at index 2 of miRNA and U belongs to the target sequence. $i\_MisMatch$ shows the nucleotide at position i does not bind to the target. Features $i\_Bulge\_on\_miRNA$ or $i\_Bulge\_on\_target$ represent a bulge at index i on miRNA or target respectively. We then ran CFS on each of five subsets provided by the MHL method. The extracted features by CFS for the subsets 1 to 5 are shown in Table 3 in columns two to six. By contrasting columns two to six versus column one we can see several biological details that are missing in column one; The appearance of adenine in the first position of target, i.e., $t1\_A$, is a major identifier of targets for many human miRNAs [25] and MHL assigned all such samples to subset 2. This $t1\_A$ is missing in column one which indicates CFS was not able to extract

Table 3. Feature selected by CFS (Correlation based Feature Selection), from complete training set versus from five subsets partitioned by the MHL algorithm. The comparison of first row with rows two to six shows that MHL can help to extract biological details from subsets while they could not be captured by the CFS method on the complete training set. In each feature the number represents the nucleotide index starting from 1 on 3' side of miRNA. For example, 2_AU means Adenine in the second position of miRNA binds to a Uracil on target.

All Training Data	Subset 1	Subset 2	Subset 3	Subset 4	Subset 5
2_AU	2_AU	t1_A	2_AU	2_AU	2_AU
2_UA	2_UA	2_AU	2_UA	2_MisMatch	2_UA
2_MisMatch	2_GC	2_UA	3_UA	4_AU	2_GC
3_UA	3_UA	2_GC	4_AU	7_GU	3_UA
4_AU	4_AU	2_MisMatch	4_Bulge_on_miRNA	8_Bulge_on_target	4_AU
4_UA	4_UA	2_Bulge_on_miRNA	5_AU	19_UG	4_UA
4_Bulge_on_miRNA	4_Bulge_on_miRNA	3_UA	6_AU	20_AU	4_Bulge_on_miRNA
5_AU	5_AU	3_Bulge_on_miRNA	6_UA	20_Bulge_on_target	5_AU
5_UA	5_UA	4_AU	7_AU	22_AU	5_UA
6_AU	5_Bulge_on_miRNA	4_UA	8_AU	MFE	6_AU
6_Bulge_on_miRNA	6_AU	5_UA	8_UA		6_Bulge_on_miRNA
7_AU	6_Bulge_on_miRNA	6_AU	9_Bulge_on_target		7_AU
8_AU	7_AU	6_GU	10_Bulge_on_target		8_AU
8_UA	7_Bulge_on_miRNA	7_AU	15_AU		8_UA
10_Bulge_on_target	8_AU	8_AU	21_UA		10_Bulge_on_target
15_AU	8_UA	8_UA	22_AU		16_AU
20_AU	13_AU	9_GU	22_GU		20_AU
22_GU	13_GU	10_GU	MFE		20_Bulge_on_target
MFE	15_AU	16_AU			21_UA
	19_UA	17_UA			MFE
	20_Bulge_on_target	21_AU			
	21_GU	21_UA			
	22_AU	22_AU			
	MFE	MFE			

it from the complete training set. There are several confirmed miRNAs with G:U base-pairs or single nucleotide bulges in their seeds [5]. CFS on all training data only has been able to detect bulges in positions 4 and 6 on miRNA, but with MHL we can confirm these bulges in indexes 2, 3, 5 and 7, in addition to the 4 and 6 positions. Subset 2 includes miRNAs with bulges in indexes 2 and 3 and subset 4 contains miRNAs with no bulge in seed area. Subset 1 has miR-NAs with bulges in the rear half of the seed, positions 4, 5 and 7. G:U wobble base pairs have not been recognized in the first column (CFS without MHL), but with MHL miRNAs with G:U base pairs have been detected and grouped into subset 4. GC is a strong base-pair and a proposed biological mechanism of targeting [32] claims base-pairs on positions 2, 3 and 4 make the groove inside the Argonaut protein open to accommodate the target. MHL has been able to cluster samples related to this mechanism into subsets 1, 2 and 5. Column one, features extracted from all training data, does not include any GC base-pair in miRNA target duplexes. Pairing to the 3' side of miRNA can compensate for single-nucleotide bulges or mismatches in the seed region [2]. The first column does not show a significant presence of such pairs, but columns 1 to 4 have contiguous base-pairs and also have more individual pairs at the 3' side of miRNA; Subsets 1 to 4 contain samples with adjacent pairs at positions 19 to 22. Subset 2, in addition, contains two more contiguous base pairs at indexes 16 and 17. The feature 2_mismatch separates canonical seed samples from non-canonicals, and MHL partitioned these two main type of samples into subsets 1, 3 and 5 for canonicals, versus 2 and 4 for non-canonicals. Splitting samples into canonical versus non-canonical subsets has not been explicitly coded into the MHL, but MHL has been able to automatically learn exclusive hypotheses for them and cluster the data accordingly.

These biologically interpretable details seen in subsets 1 to 5 could not be extracted by the same feature selection algorithm on the complete dataset; in other words the first column of the table lacks the details. This shows that the MHL algorithm may provide subsets of the data that have biologically correlated samples. The subsets can be further studied to determine targeting mechanism of each sample and therefore for the associated miRNA. Based on the current understanding of miRNA targeting mechanisms some subsets or features may not have a known biological interpretation, but they can be used in *in vivo* experiments to discover and verify new targeting mechanisms.

A unique advantage of MHL versus standard clustering algorithms is that MHL clusters the data based on the optimum hypotheses it learns. To compare the clustering performance of MHL with other popular algorithms, we ran several clustering algorithms from Weka package - Canopy, Cobweb, EM (Expected Maximization), FarthestFirst, FilteredClusterer and Xmeans [38] on our dataset. The results are shown in Table 4; column one lists algorithms and the second column is the number of clusters created by each algorithm. These methods either created too many clusters or split the dataset into two large subsets.

Table 4. This table compares MHL clustering performance versus other standard and popular algorithms. We ran these methods from the Weka package, listed in the first column, on our dataset. The second column shows the number of clusters created by each algorithm. These algorithms either created too many clusters or split the dataset into two large subsets. To evaluate the relevance of samples to each other in a cluster, we then ran the CFS algorithm. If samples in a cluster are relevant, the CFS algorithm gives a high *Merit score* for the selected features in the cluster. The *Merit score* is between 0 and 1. For each algorithm, we computed the weighted average of *Merit scores* for the clusters created by the algorithm, and it is shown in the column three. By comparing columns two and three for other algorithms versus MHL, one can see MHL could give better clusters in terms of the number of subsets and the average *Merit score* for the subsets.

Clustering algorithm	Number of clusters	*Merit scores* (weighted average of clusters)
Canopy	100	0.629
Cobweb	917	0.016
EM	15	0.441
FarthestFirst	2	0.56
FilteredClusterer	2	0.51
Xmeans	2	0.51
MHL	5	0.548

If samples in a cluster are relevant, the CFS algorithm gives a higher *Merit score* for the selected features than if the samples are not related. The *Merit score* is between 0 and 1; a high *Merit score* means a low correlation between the selected features and a high correlation to the sample label. For each cluster provided by the algorithms, we ran CFS and used the *Merit score* as an estimate for the relevance of samples in the cluster. As the number of clusters provided by different clustering algorithms varies, we computed the weighted average of *Merit scores* for each algorithm as shown in the column three of Table 4. Weight for each score is the proportion of the cluster size to the total number of samples in the training dataset. Sum of the weights for each clustering algorithm is one.

Table 4 shows that MHL does better clustering in terms of the number of clusters and the average *Merit score* for the dataset. MHL created five clusters with an average *Merit score* of 0.548. The only algorithms with better *Merit scores* are *FarthestFirst* with the score 0.56 and *Canopy*'s average score is 0.629. *FarthestFirst* has an tiny advantage for the *Merit score* but it created two large clusters. *Canopy* has the highest score while it created 100 small clusters. MHL is nicely in balance for both the number of clusters and the average *Merit score* and that is because of our unique approach for clustering, i.e., clustering by the learned hypotheses. For each hypothesis learned, MHL chooses those samples that the hypothesis could either accept or reject with confidence, a vote close to one or zero respectively. Other algorithms, however, partition the dataset based on attributes similarity. Sample attributes have different correlations to the label, and this could mislead the clustering algorithm away from partitioning by the most predictive attributes. MHL exhaustively searches for the hypotheses with the highest predictive performance and eventually those optimum hypotheses cluster the dataset; thus MHL created clusters are biologically meaningful and lead to better predictive performance.

Given the importance of miRNA target prediction, dozens of algorithms has been published in the last two decades. These methods vary based on the information they use, their accessibility, being rule-based or data-driven and their fundamentals. For a fair comparison of our method versus state-of-the-art methods, we studied several of the renown tools, such as, TargetScan [1], TarPmiR [7], miRBase [13], DIANA-microT [26], miRanda [4], miTarget [18], RNAhybrid [30], Avishkar [12], TargetSpy [33], miRWalk [8] and miRanda-mirSVR [3]. Some of these methods utilize other information, for example, TargetScan [1], miRanda-mirSVR [3] and DIANA-microT [26] use sequence conservation across species, conserved or non-conserved microRNA family, and miRBase annotation. Our MHL target predictions rely only on sequences of the miRNA and its target site. Comparing MHL with methods like TargetScan would be technically not feasible because our collected data do not have the information used by TargetScan. Source code or executable files for some methods are not available, and they are accessible through their online web sites, for instance, miRDB [36], miTarget [18] and miRWalk [8]. Our test set composed of 6,646 samples (HSA 6,129 samples and MMU 517) and we could not submit the test set as thousands of online queries manually. Functionality of some of these methods is different than our method, for example, TarPmiR [7] finds targets across a given mRNA for a miRNA sequence while our data samples are pairs of short miRNA and target site sequences, and MHL is about finding out if a pair bind to each other. From available software and methods for miRNA tareget prediction, we could only use those with downloadable source code or executable code. Moreover, for a fair comparison, such methods would also need to predict a target site, merely based on sequences of miRNA and target site. Software tools satisfying all requirements were miRanda [4] and RNAhybrid [30]. These methods rely on fundamental principles of miRNA targeting mechanism such as a lower free energy binding and a stable secondary structure duplex. These metrics have had reliable performance, therefore miRanda and RNAhybrid are still widely used either solely or as core components of other algorithms such as miRanda-mirSVR [3] and miRanda-MiRBase [27].

In the previous work by the authors [28] on the same dataset, RNAhybrid and miRanda performed similarly and with Area Under the Cure (AUC) of 0.5. For miRanda, $miRanda\ Score$ and $miRanda\ MFE$, and for RNAhybrid $RNAhybrid\ MFE$ were used to distinguish targets from non-targets. In terms of AUC, MHL surpasses these methods with high margins and there are two main reasons for that; first RNAhybrid and miRanda are rule-based while MHL is data-driven, and second MHL hypotheses are trained on experimentally verified samples, i.e. on mirTarBase records. miRNAs duplexes with $non-canonical$ seeds [23] might have higher MFE than duplexes with $canonical$ seeds, due to mismatches in the seed area. 58% of mirTarBase are $non-canonicals$ and MFE based target prediction tools such as RNAhybrid and miRanda may not be able to detect them. On the other side, MHL learned specific hypotheses for $non-canonical$ samples, and this gives a unique advantage to our algorithm to outperform rule-based methods.

5   Conclusion

miRNAs are small endogenous non-coding RNA molecules that have a critical function in suppressing genes and their dysfunction has been associated with many diseases including cancer. Due to the importance of their effects in several cell mechanisms, biologists are very interested to discover their functionality. Their function may be correlated with the way they recognize their targets. A lot of research has been ongoing to develop algorithms for miRNAs' target prediction. In this work, we presented a multi hypotheses learner algorithm (MHL) that aims for two purposes; first to predict miRNA targets with a high accuracy and second to provide partitions of samples biologically correlating with each other in a partition. These partitions can potentially be used for better understanding of targeting mechanisms as well as providing sequences for *in vivo* experiments, to discover new mechanisms.

Our evaluations and results show that the partitioning approach can significantly improve the performance of a machine learning method. Moreover, feature selection in the resulting partitions reveals that the MHL partitioning mechanism is indeed biologically meaningful and partitions have exclusive and distinctive features that are confirmed in the biology literature.

Acknowledgement. The authors would like to thank Dr. Cory Momany and Dr. Khaled Rasheed for their feedbacks and comments on this work.

References

1. Agarwal, V., Bell, G.W., Nam, J.W., Bartel, D.P.: Predicting effective microrna target sites in mammalian mRNAs. Elife **4**, e05005 (2015)
2. Bartel, D.P.: MicroRNAs: target recognition and regulatory functions. Cell **136**(2), 215–233 (2009)
3. Betel, D., Koppal, A., Agius, P., Sander, C., Leslie, C.: Comprehensive modeling of microrna targets predicts functional non-conserved and non-canonical sites. Genome Biol. **11**(8), R90 (2010)
4. Betel, D., Wilson, M., Gabow, A., Marks, D.S., Sander, C.: The microRNA. org resource: targets and expression. Nucleic Acids Res. **36**(suppl 1), D149–D153 (2008)
5. Brennecke, J., Stark, A., Russell, R.B., Cohen, S.M.: Principles of microrna-target recognition. PLoS Biol. **3**(3), e85 (2005)
6. Cloonan, N.: Re-thinking miRNA-mRNA interactions: intertwining issues confound target discovery. BioEssays **37**(4), 379–388 (2015)
7. Ding, J., Li, X., Hu, H.: TarPmiR: a new approach for microrna target site prediction. Bioinformatics **32**(18), 2768–2775 (2016)

8. Dweep, H., Sticht, C., Pandey, P., Gretz, N.: miRWalk-database: prediction of possible miRNA binding sites by "walking" the genes of three genomes. J. Biomed. Inf. **44**(5), 839–847 (2011)

9. Enright, A.J., et al.: MicroRNA targets in drosophila. Genome Biol. **5**(1), R1–R1 (2004)

10. Friedman, R.C., Farh, K.K.H., Burge, C.B., Bartel, D.P.: Most mammalian mRNAs are conserved targets of micrornas. Genome Res. **19**(1), 92–105 (2009)

11. Genome.gov. https://www.genome.gov/10001345/. Accessed 06 Jan 2017

12. Ghoshal, A., Grama, A., Bagchi, S., Chaterji, S.: An ensemble SVM model for the accurate prediction of non-canonical microRNA targets. In: Proceedings of the 6th ACM Conference on Bioinformatics, Computational Biology and Health Informatics, pp. 403–412. ACM (2015)

13. Griffiths-Jones, S., Grocock, R.J., Van Dongen, S., Bateman, A., Enright, A.J.: miRBase: microRNA sequences, targets and gene nomenclature. Nucleic Acids Res. **34**(suppl-1), D140–D144 (2006)

14. Hall, M.A.: Correlation-based feature selection for machine learning. Ph.D. thesis, The University of Waikato (1999)

15. Hsu, S.D., et al.: miRtarBase update 2014: an information resource for experimentally validated mirna-target interactions. Nucleic Acids Res. **42**(D1), D78–D85 (2014)

16. Jansson, M.D., Lund, A.H.: MicroRNA and cancer. Mol. Oncol. **6**(6), 590–610 (2012)

17. John, B., Enright, A.J., Aravin, A., Tuschl, T., Sander, C., Marks, D.S.: Human microrna targets. PLoS Biol. **2**(11), e363 (2004)

18. Kim, S.K., Nam, J.W., Rhee, J.K., Lee, W.J., Zhang, B.T.: miTarget: microRNA target gene prediction using a support vector machine. BMC Bioinf. **7**(1), 411 (2006)

19. Kiriakidou, M., et al.: A combined computational-experimental approach predicts human microrna targets. Genes Dev. **18**(10), 1165–1178 (2004)

20. Lewis, B.P., Burge, C.B., Bartel, D.P.: Conserved seed pairing, often flanked by adenosines, indicates that thousands of human genes are microRNA targets. Cell **120**(1), 15–20 (2005)

21. Lewis, B.P., Shih, I.H., Jones-Rhoades, M.W., Bartel, D.P., Burge, C.B.: Prediction of mammalian microRNA targets. Cell **115**(7), 787–798 (2003)

22. Lin, S., Gregory, R.I.: MicroRNA biogenesis pathways in cancer. Nat. Rev. Cancer **15**(6), 321–333 (2015)

23. Loeb, G.B., et al.: Transcriptome-wide mir-155 binding map reveals widespread noncanonical microrna targeting. Mol. Cell **48**(5), 760–770 (2012)

24. Lorenz, R., et al.: Viennarna package 2.0. Algorithms Mol. Biol. **6**(1), 26 (2011)

25. Witkos, T.M., Koscianska, E., Krzyzosiak, W.J.: Practical aspects of microRNA target prediction. Curr. Mol. Med. **11**(2), 93–109 (2011)

26. Maragkakis, M., et al.: DIANA-microT web server: elucidating microRNA functions through target prediction. Nucleic Acids Res. **37**(suppl-2), W273–W276 (2009)

27. Maziere, P., Enright, A.J.: Prediction of microrna targets. Drug Discov. Today **12**(11–12), 452–458 (2007)

28. Mohebbi, M., Ding, L., Malmberg, R.L., Momany, C., Rasheed, K., Cai, L.: Accurate prediction of human miRNA targets via graph modeling of miRNA-target duplex. J. Bioinf. Comput. Biol. (2018). https://doi.org/10.1142/S0219720018500130

29. Peterson, S.M., Thompson, J.A., Ufkin, M.L., Sathyanarayana, P., Liaw, L., Congdon, C.B.: Common features of microrna target prediction tools. Front Genet **5**, 23 (2014)
30. Rehmsmeier, M., Steffen, P., Höchsmann, M., Giegerich, R.: Fast and effective prediction of microRNA/target duplexes, vol. 10. Cold Spring Harbor Lab (2004)
31. Rusinov, V., Baev, V., Minkov, I.N., Tabler, M.: Microinspector: a web tool for detection of miRNA binding sites in an RNA sequence. Nucleic Acids Res. **33**(suppl 2), W696–W700 (2005)
32. Schirle, N.T., Sheu-Gruttadauria, J., MacRae, I.J.: Structural basis for microrna targeting. Science **346**(6209), 608–613 (2014)
33. Sturm, M., Hackenberg, M., Langenberger, D., Frishman, D.: TargetSpy: a supervised machine learning approach for microrna target prediction. BMC Bioinf. **11**(1), 1 (2010)
34. Sujit Pal, A.G.: Deep Learning with Keras. Packt Publishing, Birmingham (2017)
35. Thompson, S.K.: Sampling. Wiley, Hoboken (2012)
36. Wang, X.: miRDB: a microRNA target prediction and functional annotation database with a wiki interface. RNA **14**(6), 1012–1017 (2008)
37. Weill, N., Lisi, V., Scott, N., Dallaire, P., Pelloux, J., Major, F.: Mirbooking simulates the stoichiometric mode of action of micrornas. Nucleic Acids Res. **43**(14), 6730–6738 (2015)
38. Witten, I.H., Frank, E., Hall, M.A., Pal, C.J.: Data Mining: Practical Machine Learning Tools and Techniques. Morgan Kaufmann, Burlington (2016)
39. Yue, D., Liu, H., Huang, Y.: Survey of computational algorithms for microRNA target prediction. Curr. Genom. **10**(7), 478–492 (2009)

RiboSimR: A Tool for Simulation and Power Analysis of Ribo-seq Data

Patrick Perkins$^{(\boxtimes)}$, Anna Stepanova, Jose Alonso, and Steffen Heber

North Carolina State University, Raleigh, NC 27502, USA
Pjperki2@ncsu.com

Abstract. RNA-seq and Ribo-seq are widespread quantitative methods for assessing transcription and translation. They can be used to detect differential expression, differential translation, and differential translation efficiency between conditions. The statistical power to detect differential genes is affected by multiple factors, such as the number of replicates, sequencing depth, magnitude of differential expression and translation, distribution of gene counts, and method for estimating biological variance. As power estimation of translational efficiency involves the combination of both RNA-seq measurements and Ribo-seq measurements, this task is particularly challenging. Here we propose a power assessment tool, called RiboSimR, based purely on data simulation. RiboSimR, produces semi-parametric simulations that generate data based on real RNA and Ribo-seq experiments, with customizable choices on baseline parameters and tool configurations. We demonstrate the usefulness of our tool by simulating data based on two published Ribo-seq datasets and analyzing various aspects of experimental design.

Keywords: Ribosome profiling · RNA-seq · Simulation · Power analysis · Experimental design · Differential analysis

1 Introduction

RNA-seq and Ribo-seq are popular techniques for investigating the transcriptional and translational landscape [1, 2]. The techniques use next-generation sequencing to produce genome-wide high-resolution snapshots of the total populations of mRNAs and translating ribosomes, respectively. Although it is a less widely used assay, Ribo-seq has shown that it can be used to assess the dynamics of ribosome activity during translation at a nucleotide specific resolution [1]. These techniques generally produce count tables, which quantify transcription and translation for each gene. These tables can be used to measure changes in transcription and translation across biological conditions, treatments, or timepoints by calculating differential expression (DE), differential translation (DT), and differential translation efficiency (DTE). Differential translation measures changes in translation without taking changes in transcript abundance into account, while differential translational efficiency assesses changes in translation after accounting for changes in transcript abundance [3].

In order to determine statistical significance when testing for differential genes, sample replication is used to account for biological and technical variability [4].

© Springer Nature Switzerland AG 2020
I. Măndoiu et al. (Eds.): ICCABS 2019, LNBI 12029, pp. 121–133, 2020.
https://doi.org/10.1007/978-3-030-46165-2_10

The amount of replication researchers should employ is a complicated but important aspect of experimental design in and sequencing experiment. Additionally, for both Ribo-seq and RNA-seq, there exists a minimum sequencing depth threshold for identifying biologically important genes [5]. Sequencing under this threshold reduces the power of the test to identify differential genes, while sequencing above this threshold has diminishing returns in terms of the yield of differential genes per amount of sequencing. As we are dealing with read count data, which is commonly modeled using a negative binomial distribution, counting error plays a role in gene determining optimal sequencing depth [6, 7]. Genes with low counts are more greatly affected by this counting error than those with high counts, and therefore, our ability to detect a differential gene is dependent on the count value of the gene. Determining the appropriate levels of sample replication and sequencing depth for Ribo-seq and RNA-seq are critically important aspects of experimental design and should to be assessed prior to conducting any experiment.

Our ability to detect differential genes can also be affected by which methods we choose to use when performing tests for statistical significance. As sample sizes are often limited, empirical Bayes methods are commonly used to estimate gene-specific biological variation by shrinking variance estimates towards the average trend across all genes [6, 7]. These procedures are typically effective at providing variance estimates which are closer to the biological variance, but can also introduce dependencies amongst genes that violate some of the assumptions of the error-controlling procedures during multiple testing correction. It is essential that researchers understand the differences in methodologies between tools for differential analysis, as they often contain different variance estimation procedures and multiple testing correction methods, both of which can have significant effects on the results of the tests. Other seemingly simple decisions, like choosing a statistical significance threshold to use when determining differential genes, can be of great importance. The level of change that is considered biologically significant can vary based on the purpose of an experiment, as can trends in the relationship between p-values and effect sizes [8].

Researchers have previously shown that simulations can be used to examine different experimental design setups for RNA-seq, including scenarios with varying sequencing depth and replicate number [9–11]. Wu et al. proposed a prospective power analysis setup to visualize power in multiple forms and across various circumstances [9]. Similarly to other studies, they simulate datasets using the negative binomial distribution, and draw parameters from real experiments. In Ribo-seq, the complexity of these experimental design questions only increases, as we must ask ourselves if our decisions might effect each experiment type differently. Researchers also need to assess scenarios where there are potentially large discrepancies in the level of sampling depth and replication between RNA-seq and Ribo-seq experiments. Issues can also arise when testing for statistical significance, as evident by the increased amount of discrepancy amongst methods which employ more complex two factor tests to calculate differential translation efficiency [12–15].

We therefore propose a comprehensive simulation strategy for the purpose of performing dynamic power analyses of Ribo-seq experiments. We apply this strategy to assess various questions in experimental design of Ribo-seq and RNA-seq experiments. These questions include the optimal level of sample replication and sequencing depth, the tradeoff in value of increasing depth versus replication, the effect of a gene's expression has on our ability to identify it as differential, the effect of adding differing amounts of depth and replication to pre-existing experiments, and the effect of various tools and parameters for performing differential analysis. We show the effects of this analysis using two datasets in Arabidopsis thaliana and Toxoplasma gondii. Furthermore, we present a R Shiny web application which allows users to perform simulations and power analyses based on their own data or pilot datasets. This tool can be used to decide which experimental parameters a user wants to use in a prospective project, or to aid them to add to a preexisting study.

2 Methods

We propose the web application RiboSimR, which uses simulations to evaluate the effect of experimental design on our power to identify differential genes in Ribo-seq experiments. Our simulation methodology is largely adapted from Wu et al.'s work on RNA-seq simulation, and our tool is composed of two similar steps [9]. Initially, we use a semi-parametric simulation scheme to produce count tables for Ribo-seq and RNA-seq experiments. These tables are made by drawing parameters from experimental count tables provided within the tool, or optionally uploaded by the user. The simulations therefore borrow multiple aspects from the real data, such as the distribution of biological dispersion within genes, the distribution of average transcription and translation levels, the negative relationship between dispersion and count magnitude, and the positive association between average transcripts and footprints. Secondly, we assess power and false discovery rates (FDRs) for the simulations using a large number of custom visualizations. Within the output of RiboSimR, we highlight features like false discovery cost, stratified power, and sequencing value. The tool allows users to customize various parameters within both the simulation setup and output generation steps.

2.1 Simulation Strategy

In order to perform downstream power analyses, we first use a negative binomial model to simulate both RNA-seq and Ribo-seq count tables. Researchers commonly use negative binomial models to generate RNA-seq and Ribo-seq count data, because the flexibility of these models allows for accurate representation of the mean-variance relationship found in these data [6, 7]. The NB distribution corresponds to a gamma-Poisson mixture, with the gamma distribution representing the biological variation, and the Poisson layer modeling the variability in sequencing read counts [9]. If Y_{gi} is the count value for gene g in replicate i, then $Y_{gi} \sim NB(s_i \mu_g, \phi_g)$, where μ_g is the mean count for gene g, ϕ_g is the dispersion for gene g, and s_i is the normalization factor for the library size of replicate i. We use this technique to model RNA-seq and Ribo-seq counts separately, i.e. each gene has a different mean count for Ribo-seq (Y_{giRibo}) and RNA-seq (Y_{giRNA}). Our method uses non-parametric resampling of mean count and dispersion parameters empirically from existing datasets, as we lack a valid justification for drawing the parameters parametrically. The dispersion parameter corresponds to the squared biological coefficient of variance, and can be thought of as representing the biological variation of gene expression between replicates [6, 7]. We choose to sample mean count and dispersion parameters for individual genes as pairs in order to preserve their relationship, which has been previously described [6, 7]. Additionally, we sample parameters for our simulated RNA-seq and Ribo-seq counts in pairs, as there is a positive relationship between the level of transcription and translation within genes. For example, for a simulated gene g, we will sample parameters $[\mu_{gRNA}, \Phi_{gRNA}, \mu_{gRibo}, \Phi_{gRibo}]$ from a single row of the empirical count table.

Another important step in our simulations is to set effect sizes. As we seldom know the precise amount of DE, DT, and DTE that is biological significant within an experiment, this can be a risky assumption to make. These effect sizes are applied to mean count parameters across the two conditions, in effect generating artificial differential genes. For Ribo-seq experiments, we must assume separate effect sizes across the two different experiment types. Several techniques have been used to set effect sizes for simulated RNA-seq experiments, but we will focus on using a mixture model [9–11]. For a one factor test, such as a test for DE in RNA-seq, we let zg be the indicator that gene g is differentially expressed across a given treatment, and the proportion of differential genes be $P(z_g = 1) = \pi_1$. We have the effect size β_g satisfying $\beta_g \mid z_g = 0 = 0$ and $\beta_g \mid z_g = 1 \sim N(0,\sigma^2)$. This would be designated as a zero-inflated normal distribution for β_g. For Ribo-seq, because we are dealing with two experimental types, we require both $\beta_{gRibo}, \beta_{gRNA}$ and z_{gDT}, z_{gDE}, which describe the effect size and differential indicators for changes across conditions in either experiment type. We also define the differential effect size between experiments as β_{gE} where $\beta_{gE} = |\beta_{gRibo} - \beta_{gRNA}|$, and z_{gDTE} as the indicator for differential translation efficiency. Therefore, we have four possible types of differential genes:

1. $\beta_{gRNA} \neq 0, \beta_{gRibo} = 0, \beta_{gE} \neq 0 \rightarrow z_{gDE} = 1, z_{gDT} = 0, z_{gDTE} = 1$
2. $\beta_{gRNA} = 0, \beta_{gRibo} \neq 0, \beta_{gE} \neq 0 \rightarrow z_{gDE} = 0, z_{gDT} = 1, z_{gDTE} = 1$
3. $\beta_{gRNA} \neq 0, \beta_{gRibo} \neq 0, \beta_{gE} = 0 \rightarrow z_{gDE} = 1, z_{gDT} = 1, z_{gDTE} = 0$
4. $\beta_{gRNA} \neq 0, \beta_{gRibo} \neq 0, \beta_{gE} \neq 0 \rightarrow z_{gDE} = 1, z_{gDT} = 1, z_{gDTE} = 1$

Up to this point, we have described genes with zg = 1 as differentially expressed, but it may be the case that these genes are not biologically interesting, as $|\beta_{gRNA}|$ or $|\beta_{gRibo}|$ may be extremely low but non-zero. We would expect to have little power to detect these genes. Thus, we may be interested in defining differential genes of interest with an indicator zg* = 1 if $|\beta_g| \geq \Delta$. This allows us to investigate the power of finding genes which we think are more likely to be biologically relevant. We can let the user decide the 'meaningful effect size', Δ. The meaningful effect size can also be different between experiment types, as users may expect larger overall differences to occur in either RNA-seq or Ribo-seq, or want to relax constraints on either experiment type.

2.2 Differential Analysis

Once the count tables for both experiment types have been simulated, we can assess differential expression, differential translation, and differential translation efficiency. Table 1 depicts the scenarios for each type of differential test, assuming a meaningful effect size. We have implemented four tools for analysis of differential genes: edgeR,

DESeq 2, limma-voom, and DSS for analysis of DE, DT, and DTE [6, 7, 17, 18]. Tests for DE and DT essentially ignore counts from the opposite experiment type, while tests for DTE perform a two factor test which investigates changes in translation levels after accounting for changes in levels of transcription. This can be seen as conducting an overall test for interaction using the formula~Experiment + Condition + Experiment: Condition, where we are looking for genes which respond differently to the treatment relative to the experiment type [6].

Table 1. Differential scenarios, assuming effect size Δ

	$\|\beta_{gRibo}\| < \Delta$		$\|\beta_{gRibo}\| \geq \Delta$	
	$\|\beta_{gE}\| < \Delta$	$\|\beta_{gE}\| \geq \Delta$	$\|\beta_{gE}\| < \Delta$	$\|\beta_{gE}\| \geq \Delta$
$\|\beta_{gRNA}\| < \Delta$	$z^*_{gDTE} = 0$	$z^*_{gDTE} = 1$	$z^*_{gDTE} = 0$	$z^*_{gDTE} = 1$
	$z^*_{gDE} = 0$	$z^*_{gDE} = 0$	$z^*_{gDE} = 0$	$z^*_{gDE} = 0$
	$z^*_{gDT} = 0$	$z^*_{gDT} = 0$	$z^*_{gDT} = 1$	$z^*_{gDT} = 1$
$\|\beta_{gRNA}\| \geq \Delta$	$z^*_{gDTE} = 0$	$z^*_{gDTE} = 1$	$z^*_{gDTE} = 0$	$z^*_{gDTE} = 1$
	$z^*_{gDE} = 1$	$z^*_{gDE} = 1$	$z^*_{gDE} = 1$	$z^*_{gDE} = 1$
	$z^*_{gDT} = 0$	$z^*_{gDT} = 0$	$z^*_{gDT} = 1$	$z^*_{gDT} = 1$

2.3 Power Assessment

For our purposes, we are not interested in finding genes with variation between conditions which is less than the effect size Δ, and can therefore restrict our genes to two categories: $\|\beta_g\| < \Delta$ and $\|\beta_g\| \geq \Delta$, where βg represents the effect size for the test of interest, be it DE, DT, or DTE. Table 2, which has been adopted from Wu et al. and simplified, shows the possible scenarios for any test [9].

Let G be the total number of genes tested, where G_1 are genes which exhibit an effect size of at least Δ, and G_0 are genes which do not. D_g is the decision on any gene, where $D_g = 1$ is a discovery and $D_g = 0$ is a non-discovery. Here V represents the total number of type I errors, or false discoveries, S the number of true positive, and R the total number of discoveries. The type 1 error is therefore $P(V > 0)$ and the FDR is $E[V/R]$. The FDC, described by Wu et al., can be defined as the number of false discoveries made for every true discovery, and represented as $E[V/S]$ [9].

Here we are labeling genes which have $0 < \|\beta_g\| < \Delta$ as false discoveries, meaning that we only want to measure our ability to find genes with a meaningful effects size, and penalize all other discoveries.

Therefore, when we talk about power, we are actually discussing a targeted power, instead of a family-wise power, which is the probability of detecting all true differential genes [8]. This is because we both doubt our ability to find genes with very small effect sizes, and also question the biological importance of such genes. The targeted power, which from now on we simply refer to as power, is therefore $E[S/G_1]$. For each simulation, we calculate the power, false discovery cost, and true positive rate of the discoveries made using a differential analysis tool and nominal FDR provided by the user. Averages across a number of repeated simulations are reported.

Table 2. Differential scenarios, assuming Δ

	Differential?	Discovery?		Total
	z_g^*	$D_g = 1$	$D_g = 0$	
$|\beta_g| < \Delta$	0	V	$G_0 - V$	G_0
$|\beta_g| \geq \Delta$	1	S	$G_1 - S$	G_1
Total		R	$G - R$	G

2.4 Implementation

We have implemented the proposed methods for simulation and power analysis in a R Shiny web application called RiboSimR, available at http://pjperki2.shinyapps.io/power. The tool allows users to upload their own count files for RNA-seq and Ribo-seq and create simulated data sets based on the parameters of their experiments. Users may also test different experimental scenarios for prospective experiments using provided pilot data sets. The app allows multiple options to customize the simulations, by choosing the size of the datasets they wish to simulate, the number of repetitions, and the meaningful effect size. Users may also choose between a large number of different visualizations, which include simulated variations of sample replication and sequencing depth, gene count stratification, comparisons between differential analysis tools, etc.... The tool also provides the choice of which type of test they want to conduct (DE, DT, DTE), and which power metrics they want to assess (power, true positive rate, false discovery cost). A description of the various tools and functionalities included in RiboSimR can be found at the apps home page. Runtime for the simulations depends on the size of the simulated data and the number of repetitions conducted, but a typical simulation with 20,000 genes and 20 repetitions takes approximately 10–30 min.

3 Results

3.1 Data Preparation

To display the utility of RiboSimR, we performed simulations and power analyses using two published datasets. The Merchante et al. data is from Arabidopsis thaliana seedlings, and was used to study of the role of the phytohormone ethylene as a gene-specific regulator of translation [19]. The data are split into two conditions, control samples treated with air, and ethylene treated seedlings. The Hassan et al. data are from intracellular and extracellular Toxoplasma gondii parasites [20]. This data was used to quantify the differences in translation between the intracellular and extracellular stages of the parasites.

Both datasets contain two replicates from two conditions for both RNA-seq and Ribo-seq. These datasets were chosen because they originate from studies which focus on using both RNA-seq and Ribo-seq data to identify genes which show differential translation efficiency between two treatments, and because they are from different organisms, and show varying levels of biological variability.

For all of the following simulations, we generate 20,000 genes and repeat each of the simulations 20 times. We assume that 2% of all genes fall into each of the four types of differential genes described in Sect. 2.1. Therefore, accounting for overlap, 6% of the total set of genes are DE, 6% are DT, and 6% are DTE. These simulation parameters were chosen to mirror the conditions of similar RNA-seq experiments [9]. The effect size for non-differential genes is set to 0, while the effect sizes of DE and DT genes are drawn from a normal distribution~$N(0, 1.52)$. Means and dispersions for simulated genes are drawn from the real data in paired sets with dependency. Unless stated otherwise, we employ the quasi-likelihood F-test in edgeR to calculate genes with DE, DT, and DTE. A meaningful effect size of 0.5 is used to determine biologically meaningful genes. All of the following results and figures represent output from riboSimR.

3.2 Simulation Results

Figure 1 examines how the power of our tests for DTE, DE and DT are affected by artificially altering the number of sample replicates and the sequencing depth. The x-axis represents the factor by which the empirical depth is multiplied. For DTE and DT, it seems as though, for a given replicate number, sequencing depth increases cease to yield increases in power at around 75% of the real dataset size. This indicates that increasing sequencing depth beyond this point would have diminishing returns.

However, the power to identify DE genes continues to increase as sequencing depth increases. For each type of test, it is clear that higher power can be achieved by increasing the number of sample replicates. This increase in power does not seem to have diminishing returns, as increases are observed in up to ten replicates. Our power to identify genes which show DE seems more largely reliant on both sequencing depth and replicate number than tests for DT and DTE.

To further quantify the differing effects of increasing the sequencing depth and replicate number, we can compare experiments in which the same number of total reads are added, but added by increasing only either depth or replicate number. Figure 2A shows the results of such an analysis, where the blue line indicates increasing the total number of reads by increasing depth, and the orange line indicates increasing the total number of reads by the same amount, but via increased replication. These results once again highlight that, for the Merchante dataset, power can only be increased by adding replicates.

We also investigated increasing replication and depth at differing levels in the two experiment types. The x-axis of Fig. 2B represents increasing levels of Ribo-seq replicates, while the different colored lines represent the different numbers of RNA-seq replicates. These results indicate that increasing the number of RNA-seq replicates generally has a larger positive effect on power than increasing Ribo-seq replicates. Figure 2C shows the result of a similar experiment, but increasing depth instead of replication. Here

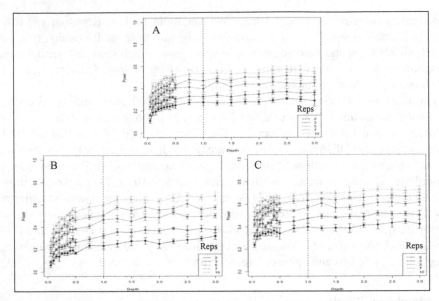

Fig. 1. Simulations based on Merchante et al. data for (A) DTE, (B) DE, and (C) DT. X-axis represents the factor by which depth is changed relative to the complete dataset. Different colored lines represent number of simulated replicates. (Color figure online)

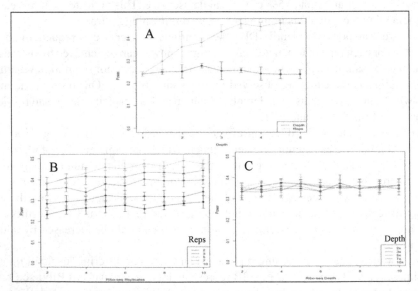

Fig. 2. (A) Value comparison between increasing reads via replications and sequencing depth. (B) Effect of increasing replication at differing levels across experiment types. (C) Effect of increasing sequencing depth at differing levels across experiment types. (Color figure online)

we are simulating the addition of reads to existing datasets, and therefore we begin simulating at 200% of empirical read levels. As previously noted, adding additional depth to
the Merchante dataset has proven to have little effect on power, and these results indicate
that addition of varying levels of RNA-seq or Ribo-seq have little effect on these results.

Figure 3 shows a quantification of the power to identify genes which show DTE,
stratified by average RNA-seq read (3A) count and Ribo-seq read count (3B). These
results indicate that, for a given replicate number, our power to detect genes with DTE
is largely reliant on both the RNA-seq and Ribo-seq count for the genes. Our power to
predict genes with lower than 50 average reads across replicates is significantly lower.
For the RNA-seq reads, the difference in power between different replicate values is
more pronounced in low read count genes, indicating that increasing the number of
replicates can have a significant effect on our ability to find low read count genes which
truly show DTE.

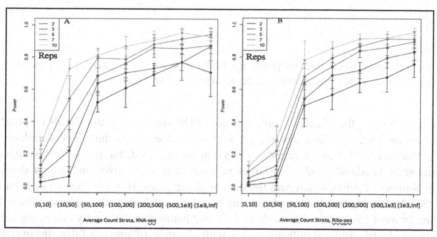

Fig. 3. Power stratification for DTE by average count value in (A) RNA-seq and (B) Ribo-seq.
Simulated based on data from Hassan et al.

We also look to measure the effects that our choice of tools and parameters have when
testing for differential genes. Figure 4 shows the use of four different tools, edgeR,
DESeq 2, limma-voom, and DSS, for detection of DTE. The tools are evaluated by
measuring power and FDR from simulations based on both the Merchante and Hassan
datasets. The simulations are performed for both 2 and 5 replicates. The results in Fig. 4A
and C indicate that DESeq 2 and edgeR generally achieve the highest power for each
of the experimental setups, while limma-voom and DSS yield lower power. Figures 4B
and D show that, in terms of FDC, limma-voom appears to outperform DESeq 2 and
edgeR for both datasets, while DSS performs well for the Hassan data and poorly for the
Merchante data. These results indicate that nuanced differences between program may
effect results differently on an experiment-by-experiment basis. It is therefore valuable
to have a way which we can evaluate each program for a user's specified dataset.

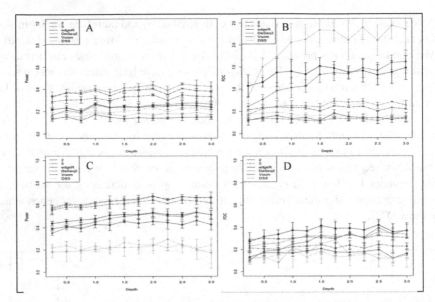

Fig. 4. Tool comparison results for power and FDC of (A, B) Merchante et al. data and (C, D) Hassan et al. data. Line types differentiate between using 2 and 5 replicates.

Figure 5 show the effects of three different FDR significance thresholds on power, true positive rate, and false discovery cost for the Merchante data. The simulations are performed for 1,5 and 8 replicates. As can be expected, the use of less stringent significance thresholds leads to a larger number of true positives and higher power, as the number of genes which pass the threshold increases. However, as can be seen in Fig. 5C, the false discovery cost when using less stringent thresholds can be significantly higher. In most cases, researchers should choose parameters values by balancing their ability to identify truly significant genes with the cost of making false discoveries. Interestingly, these results also depict a relationship between the significant threshold, replicate number, and FDC. The differences in power and FDC between the three FDR thresholds changes seems to be more drastic for two replicates than five and eight. This implies that the choice of FDR threshold has a larger effect on false discovery for experiments with a limited number of replicates.

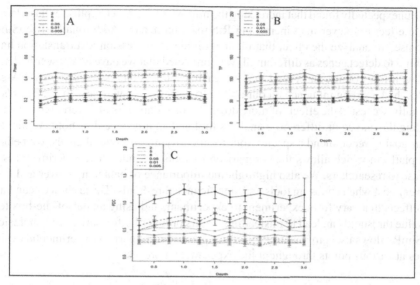

Fig. 5. The effect of using different FDR significance thresholds on (A) power, (B) true positive rate, and (C) false discovery cost. Colors represent differences in number of replicates. (Color figure online)

4 Conclusion

In this paper, we describe RiboSimR, an R Shiny web application for the analysis of experimental design parameters in Ribo-seq and RNA-seq experiments. This tool can be used prior to performing sequencing in order to test experimental design parameters using provided pilot data, or after the fact, to test the effects of adding additional sequencing information to an existing study. Using two published Ribo-seq and RNA-seq datasets, we demonstrate how RiboSimR can be used to investigate the effects of these various factors on the power to detect genes which exhibit differential expression, differential translation, and differential translation efficiency.

Firstly we showed that we can assess the effects that variations in replicate number and sequencing depth have on power to detect differential genes. We provided an example of an experiment which has reached a critical sequencing depth threshold in terms of identifying DTE and DT genes. For researchers, having the ability to predict the effects of increasing sequencing depth can save them valuable time and money. The results from our simulations also confirm the previously asserted notion that increasing sample replication yields larger increases in power than increasing sequencing depth.

We further looked and quantify the effects of increasing replication and depth differently between experiment types. This simulation setup can aid researchers in testing the experimental design parameters of an experiment they have already conducted, and help them predict how recreating the experiment with different factors, or adding additional information to their experiment, might affect their results.

We unexpectedly found that increasing the number of RNA-seq replicates has a larger positive effect on power than increasing Ribo-seq replicates. Additionally, RiboSimR can be used to analyze the effect that the magnitude of expression and translation have our ability to detect genes as differential. We confirmed that we have less power to detect genes with low counts, and found that increasing replication, especially for RNA-seq samples, had a strong influence on our ability to detect genes with less than 50 reads.

Finally, we tested the effects of more downstream parameters of differential analysis, including the choice of differential analysis tool and significance threshold. While it is not our goal to recommend a specific tool for identifying differential genes, we believe that a platform which allows the comparison of tools for individual experiments is of great use to researchers. We also highlight the importance of balancing power and false discovery cost when choosing tools and significance thresholds. These choices can have large effects that vary from experiment to experiment, and using an out-of-the-box tool or p-value threshold can lead to misleading results or loss of information. In conclusion, RiboSimR allows users to quantify the consequences of important experimental design choices at various points throughout the experimental process.

References

1. Ingolia, N., Ghaemmaghami, S., Newman, J., Weissman, J.: Genome-wide analysis in vivo of translation with nucleotide resolution using ribosome profiling. Science **324**, 218–223 (2009)
2. Wang, Z., Gerstein, M., Snyder, M.: RNA-Seq: a revolutionary tool for transcriptomics. Nat. Rev. Genet. **10**, 57 (2009)
3. Larsson, O., Sonenberg, N., Nadon, R.: Identification of differential translation in genome wide studies. Proc. Natl. Acad. Sci. **107**, 21487–21492 (2010)
4. Hansen, K., Wu, Z., Irizarry, R., Leek, J.: Sequencing technology does not eliminate biological variability. Nat. Biotechnol. **29**, 572 (2011)
5. Robles, J., Qureshi, S., Stephen, S., Wilson, S., Burden, C., Taylor, J.: Efficient experimental design and analysis strategies for the detection of differential expression using RNA-Sequencing. BMC Genom. **13**, 484 (2012)
6. Anders, S., Huber, W.: Differential expression analysis for sequence count data. Genome Biol. **11**, R106 (2010)
7. Robinson, M., McCarthy, D., Smyth, G.: EdgeR: a Bioconductor package for differential expression analysis of digital gene expression data. Bioinformatics **26**, 139–140 (2010)
8. McCarthy, D., Smyth, G.: Testing significance relative to a fold-change threshold is a TREAT. Bioinformatics **25**, 765–771 (2009)
9. Wu, H., Wang, C., Wu, Z.: PROPER: comprehensive power evaluation for differential expression using RNA-seq. Bioinformatics **31**, 233–241 (2014)
10. Frazee, A., Jaffe, A., Langmead, B., Leek, J.: Polyester: simulating RNA-seq datasets with differential transcript expression. Bioinformatics **31**, 2778–2784 (2015)
11. Busby, M., Stewart, C., Miller, C., Grzeda, K., Marth, G.: Scotty: a web tool for designing RNA-Seq experiments to measure differential gene expression. Bioinformatics **29**, 656–657 (2013)
12. Eastman, G., Smircich, P., Sotelo-Silveira, J.: Following ribosome footprints to understand translation at a genome wide level. Comput. Struct. Biotechnol. J. **16**, 167–176 (2018)
13. Larsson, O., Sonenberg, N., Nadon, R.: Anota: Analysis of differential translation in genome-wide studies. Bioinformatics **27**, 1440–1441 (2011)

14. Olshen, A., Hsieh, A., Stumpf, C., Olshen, R., Ruggero, D., Taylor, B.: Assessing gene-level translational control from ribosome profiling. Bioinformatics **29**, 2995–3002 (2013)
15. Xiao, Z., Zou, Q., Liu, Y., Yang, X.: Genome-wide assessment of differential translations with ribosome profiling data. Nat. Commun. **7**, 11194 (2016)
16. Chang, W., Chang, J., Allaire, J., Xie, Y., McPherson, J.: Shiny: Web Application Framework for R. https://CRAN.R-project.org/package=shiny
17. Law, C., Chen, Y., Shi, W., Smyth, G.: Voom: Precision weights unlock linear model analysis tools for RNA-seq read counts. Genome Biol. **15**, R29 (2014). https://doi.org/10.1186/gb-2014-15-2-r29
18. Wu, H., Wang, C., Wu, Z.: A new shrinkage estimator for dispersion improves differential expression detection in RNA-seq data. Biostatistics **14**, 232–243 (2012)
19. Merchante, C., et al.: Gene-specific translation regulation mediated by the hormone-signaling molecule EIN2. Cell **163**, 684–697 (2015)
20. Hassan, M., Vasquez, J., Guo-Liang, C., Meissner, M., Siegel, T.: Comparative ribosome profiling uncovers a dominant role for translational control in Toxoplasma gondii. BMC Genom. **18**, 961 (2017)

Treatment Practice Analysis of Intermediate or High Risk Localized Prostate Cancer: A Multi-center Study with Veterans Health Administration Data

Khajamoinuddin Syed[1(✉)], William Sleeman IV[1,2], Joseph Nalluri[2],
Payal Soni[3], Michael Hagan[2,3], Jatinder Palta[2,3], Rishabh Kapoor[2,3],
and Preetam Ghosh[1]

[1] Department of Computer Science, Virginia Commonwealth University,
Richmond, VA 23284, USA
{lnusk,pghosh}@vcu.edu
[2] Department of Radiation Oncology, Virginia Commonwealth University,
Richmond, VA 23298, USA
{william.sleemaniv,joseph.nalluri,michael.hagan,jatinder.palta,
rishabh.kapoor}@vcuhealth.org
[3] Hunter Holmes McGuire VA Medical Center, Richmond, VA 23249, USA
payal.soni@va.gov

Abstract. Prostate cancer (PCa) is a heterogeneous disease. PCa is stratified into risk groups based on clinical factors such as T-stage, Gleason score, and baseline prostate-specific antigen. Treatments are selected based on these risk groups. However, we hypothesize that non-clinical factors such as the radiation therapy (RT) center may also impact treatment selection, and we investigate the impact of these factors on treatment selection practice and their adherence to recommended guidelines from the national comprehensive cancer network (NCCN). A total of 552 patients with intermediate or high-risk localized PCa related data was collected from 34 radiation therapy centers of the Veterans Health Administration (VHA), who were treated with definitive RT and with or without Androgen Deprivation Therapy (ADT) between 2010 and 2017. Patients' clinical information is extracted by manually reviewing their medical charts. We also extracted treatment intended and treatment administered information from consult and end-of-treatment notes, respectively. The random forest classification algorithm was used to identify the impact of clinical and non-clinical factors in treatment selection, their adherence to the treatment guidelines, and treatment alteration (i.e., change in intended and administered treatments). We created models for predicting treatment intended as well as treatment administered. Our results demonstrated that non-clinical (i.e., treatment center)

The study was funded by the Veterans Affairs (VA) National Radiation Oncology Program (NROP).

I. Măndoiu et al. (Eds.): ICCABS 2019, LNBI 12029, pp. 134–146, 2020.
https://doi.org/10.1007/978-3-030-46165-2_11

factors, along with clinical factors, are significant for predicting the adherence of treatment intended to the NCCN guidelines. Furthermore, the center served as an important factor for prescribing ADT; however, it is not associated with the duration of ADT and is weakly associated with treatment alterations. This presence of center-bias in treatment selection warrants further investigation on details of center-specific barriers for NCCN guideline adherence, and as well as the impact of center-bias on oncological outcomes.

Keywords: Prostate cancer · Radiation therapy · Treatment selection · Machine learning · Clinical informatics

1 Introduction

Prostate cancer (PCa) is the most commonly diagnosed type of cancer after breast and lung cancer. In 2018 alone, over 160,000 new prostate cancer cases and over 29,000 prostate cancer-related deaths were estimated in the United States [1]. PCa is also one of the most heterogeneous type of cancer specifically with respect to intermediate or high-risk PCa [2]. The non-invasive prostate-specific antigen (PSA) test that has led to an increase in early detection of PCa leading to more localized PCa diagnosis in recent years [3].

The National Comprehensive Cancer Network (NCCN) provides clinical practice guidelines that are created by physicians to determine the best way of treating PCa patients (besides other types of cancers), depending on their diagnosis, disease stage, age and other factors. PCa is also treated with monotherapy or polytherapy. Physicians select the treatment modality based on four major criteria - age, race, life expectancy, and NCCN Risk. Factors such as patient preferences, survivorship goals along with tumor biology also play a crucial role in optimizing the treatment modality.

A major consideration during the treatment options for PCa is to check whether the cancer is contained within the prostate gland (localized), or has spread outside the prostate (locally advanced) or has spread to other parts of the body (metastasized). Radical prostatectomy (RP), external beam radiotherapy (EBRT) and brachytherapy (BT) are the common primary treatment options for localized PCa. Hormonal therapeutics such as androgen deprivation therapy (ADT) is also used as neoadjuvant/adjuvant therapy. However, ADT as monotherapy is not recommended for intermediate and high-risk cancer patients by NCCN. Ideally, a treatment option recommendation would be based on the randomized controlled trials (RCT) that compare efficacy and morbidity of alternative treatment methods. There are no randomized trials showing that one treatment is better than the other for the above-mentioned treatment options. Hence, physicians use their personal experience and expertise to predict the outcome of these treatment methods. Physicians also tend to have difficulty weighing the relative importance of each of these factors and inherently possess biases when predicting the treatment outcomes.

Based on the aforementioned considerations, determining an optimal treatment plan for the patient can be a challenging task for the physician. In order to assist the physicians with more accurate prognosis, subsequent treatment outcome prediction, and to make informed decisions, numerous predictive tools have been developed [4]. These include probabilistic models, lookup and propensity scoring tables, risk-stratification tools, classification, and regression tree analysis, nomograms, and artificial neural networks [5,6] However, to the best of our knowledge, no models have been reported that can identify why a prescribed (or administered) treatment plan do not adhere to NCCN guidelines.

The predictive models for treatment plan (or outcome) prediction have a major disadvantage. Such models do not consider the impact of non-clinical factors associated with the treatment center. The factors associated with the treatment center have shown to play a determining role in the physicians' treatment prescription practices. Non-clinical factors can be patient-related, physician-related or practice-related. These factors include patient's preference/availability, patients' adherence, physician's availability, cost, geographical proximity, treatment centers' equipment condition/availability, treatment centers' cultural aspects, type of practice (private vs. public), availability of health resources [7–10]. However, there have not been many studies which have investigated the extent of the contribution of these factors in the treatment selection process itself. Thus the motivation of this study is two-fold:

1. To use both clinical and non-clinical features for localized and locally advanced PCa patients from multiple VHA centers and use machine learning methods to predict the treatment prescribed; such methods provide a statistical approach for calculating the weight (impact) of these clinical/non-clinical features from an empirical and retrospective point-of-view.
2. To perform quality assurance assessments across the different centers and verify if the prescribed treatments were in concordance with NCCN guidelines.

This study presents a comparative analysis of treatment prescription consistency across multiple VHA centers.

2   Materials and Methods

2.1   Data Source

The study cohort comprised of patients from the United States VHA. The VHA has 40 centers treating cancer patients with radiation therapy (RT) across the US. The patient cohort was generated as a radiation oncology practice assessment (ROPA) initiative, in which clinical data of the most recently treated 20 patients from each center was collected to assess the quality of the treatment. From here on, the generated data set is referred to as the VHA-ROPA data set. The study was approved by the clinical research ethics committee of the VHA.

2.2 Study Population

A maximum of 20 patients from 34 VHA RT centers are selected whose treatment was completed between 2010 to 2017. Patients were included if they had localized intermediate or high-risk PCa. Patients were excluded if they had previous malignancy, M1 disease, or lymph node involvement. The final cohort had 552 patients from the 34 centers with NCCN risk classified as Intermediate or High.

2.3 Definitions of Variables

Definitions of variables used in our study are as follows.

Clinical Variables: We considered pre-treatment PSA count, Gleason score (GS) [primary grade, secondary grade], Gleason Grade, Tumor staging [TNM-stage], NCCN risk group, performance status, and quality of life (QoL) measures. The values for these clinical variables were manually extracted from the consult notes.

Non-clinical Variable: We defined Center-ID as a non-clinical variable. It designates a unique ID to identify the VA radiation treatment center.

ADT Duration: NCCN guidelines define ADT duration as short term (ST) or long term (LT). ST duration is 4–6 months, and LT duration is 2–3 years. We further differentiated ADT duration based on intended and administered duration. The intended duration signifies whether it was mentioned in consult notes during treatment planning, whereas ADT administered duration is calculated based on the dates of ADT injection. Table 2 shows the ADT injection type and their effective period in months depending on the dose. Table 3 shows the distribution of ADT intended and administered duration. A third category of not otherwise specified (NOS) was used to indicate cases where ADT duration was not mentioned in consult as a treatment plan.

NCCN Concordance: We defined the treatment prescribed or administered is concordant with NCCN guidelines if they were as per the NCCN guidelines [12].

2.4 Model Selection

In this section, we present the details of feature-set selection, predictive models, machine learning algorithms, and model evaluation metrics.

We used machine learning algorithms as a statistical tool to find the association between the treatments and clinical and non-clinical features. We used a supervised machine learning algorithm called random forests (RF) [13], to find these associations. The RF algorithm, as the name suggests, is the ensemble of decision trees. The RF algorithm takes the features (clinical and non-clinical variables) and target (treatments) to build the individual trees with randomly selected uncorrelated features set. The majority target predicted from all trees becomes the final model prediction. The model also provides the significance of features in classifying the targets. The significance of all features sums to 1, where higher the significance of a feature stronger is its association with the treatments, and lower significance indicates the weaker or no association.

Feature Selection. We created two feature sets using the clinical and non-clinical features to highlight the contribution of non-clinical features. The feature sets (FS) are as below

1. FS-1: Clinical features only.
2. FS-2: Clinical and Non-clinical (Center-ID) features.

Table 1. Details of the clinical factors in the VHA ROPA dataset and their frequency distribution, NOS: Not Otherwise Specified.

Data element	Count	Percentage
Total patients	552	
Centers	34	
Gleason score Primary + Secondary	549	99.50
3 + 3	17	3.00
3 + 4	219	39.67
4 + 3	128	23.18
3 + 5	18	3.26
4 + 4	79	14.31
5 + 3	2	0.36
4 + 5	61	11.05
5 + 4	19	3.44
5 + 5	3	0.54
NOS + NOS	2	0.36
PSA	549	99.50
T Stage	549	99.50
T1a - T2a	457	82.79
T2b - T2c	64	11.59
T3a -T3b	20	3.63
TX	1	0.18
NOS	7	1.26
Risk	545	98.73
Intermediate	304	55.60
High	241	44.40
Performance Status	523	94.75
Quality of Life	400	72.46
Treatment Prescribed	552	100.0
BT	24	3.07
BT-ADT	1	0.13
EBRT	132	20.23
EBRT-ADT	382	59.28
EBRT-BT	2	0.27
EBRT-BT-ADT	11	2.00

Table 2. ADT injection effective period based on the injection type and dose

ADT injection	Dose	Effective period
Leuprolide	3.75 mg	1 month
	7.50 mg	1 month
	22.50 mg	3 months
	30.00 mg	4 month
	45.00 mg	6 months
Goserelin/Zoladex	3.60 mg	1 month
	10.80 mg	3 months

Table 3. Treatment concordance with NCCN guidelines. ST: Short Term, LT: Long Term, and NS: Not Specified

NCCN risk	Treatment	ADT Duration	Intended	Administered	Concordance with NCCN
Intermediate	ADT-BT	NS	1	–	No
		LT	–	1	Yes
	BT		24	24	Yes
	EBRT		115	115	Yes
	EBRT-ADT	LT	8	15	No
		NS	11	–	No
		ST	142	146	Yes
	EBRT-ADT-BT	ST	1	1	Yes
	EBRT-BT		2	2	Yes
High	EBRT		17	17	No
	EBRT-ADT-BT	LT	9	4	Yes
		ST	1	6	Yes
	EBRT-ADT	LT	185	145	Yes
		NS	18	–	No
		ST	12	70	No

Statistical Models. VHA-ROPA dataset has patients treated with six different treatment methods (Table 1): BT, BT-ADT, EBRT, EBRT-ADT, EBRT-BT, and EBRT-BT-ADT. Based on the available treatment plans, we built the following two models.

1. Model-1: Initial Treatment (EBRT-ADT vs EBRT): This model predicts whether the patients will be treated with EBRT and ADT (EBRT-ADT), or EBRT alone. A total of 514 patients were treated with these two techniques, among which 382 patients were treated with EBRT-ADT, and 132 patients were treated with EBRT alone.
2. Model-2: ADT Duration (EBRT-ADT-ST vs EBRT-ADT-LT): This model predicts whether the ADT prescribed duration is *short term* or *long term*. Model-2 is further divided into 2 A and 2 B. Where 2 A is EBRT with ADT intended duration and 2B is EBRT with ADT administered duration.

382 patients were treated with EBRT and ADT. Table 3 shows the treatment with intended and administered ADT duration.

These models will use machine learning techniques to serve the dual purpose of (i) creating a predictive model of initial treatment selection or ADT duration based on the clinical and non-clinical features and (ii) showing the statistical correlation of the individual features in terms of impacting the treatment selection or ADT duration process.

For each of the above mentioned models, the data set was split 80 : 20 ratio into training and test sets. We used random forest algorithm for building predictive models. Models are evaluated with macro-average precision, recall, and F1-Score.

3 Results

Here, we report the results from our proposed models. We observed that treatment non-concordance with NCCN guidelines can be due to the following two reasons:

Firstly, overall treatment may not be in concordance with NCCN guidelines. For example, high-risk cancer patients treated with EBRT alone are not in concordance with NCCN. Figure 1(A) & (B) shows the center wise all non-concordant treatment counts based on ADT intended duration (i.e., prescribed ADT) and ADT administered duration treatments respectively.

Secondly, overall treatment is in concordance with NCCN however the treatment guidelines may be partially not followed. For example, a high-risk cancer patient is treated with EBRT and ADT, but ADT duration is for short-term instead of long-term. Figure 2 (A) & (B) shows the partially non-concordant patient count of each center when patients are treated with EBRT and ADT; the counts are again based on the ADT intended and administered duration respectively.

Table 4 shows the Precision, Recall, F1-Score for model-1 (EBRT-ADT vs EBRT). The goal in this model was to classify patients with treatment intent being either EBRT or a combination of EBRT and ADT (EBRT-ADT). Model 1 with FS-2 performed better in all metrics when compared to FS-1. We observed that model-1 has F1-Score of 74% with FS-1 and 82% with FS-2. These results clearly demonstrate the significance of non-clinical feature (Center-ID) in improving the overall classification performance.

Table 4 also shows the results of model-2 (EBRT-ADT-ST vs EBRT-ADT-LT). Interestingly, in this case, FS-1 and FS-2 perform quite similarly with 94% F1-Score for models with ADT intent labels (with FS-1), while F1-Score is decreased when the ADT administered labels were used. This may mean that some external factors (not considered in our feature sets) play a role for causing the alteration from treatment from the prescribed to administered. Also, non-clinical feature (Center-ID) found to have no affect on predicting the ADT duration type as opposed to Model-1 (EBRT-ADT vs EBRT). Based on these observations, we hypothesize that while centers do play a role in determining whether to prescribe ADT or not, they do not impact the actual ADT duration, in case it was administered; in other words, all centers follow similar practice in administering ADT for localized intermediate or high-risk PCa treatment.

We next evaluated the individual significance (i.e., contributions) of each of the features from FS-1 and FS-2 in our models; the feature significance were generated using the RF algorithm. Table 5 shows the feature importance of all features in all models.

For both FS-1 and FS-2, PSA and Risk consistently ranked as significant features in all the models. Specifically, for FS-1, PSA was ranked as the top feature for Models 1, 2B. For Model-2A (ADT duration intended), Risk was ranked as the top feature. This suggests that decisions on ST or LT ADT duration depend primarily on the Risk with PSA being a secondary feature of importance; these two features are primarily responsible in deciding the ADT course at the initial treatment level; however, decisions in altering the treatment intent (as captured in Model-2B with treatment administered) are impacted by the PSA and Total Gleason score (which is the third ranked feature in this model). For Model-1, PSA was ranked as the top feature with Risk as the secondary feature and T_stage as the third significant feature suggesting that decisions on treating the patients with EBRT alone or a combination of EBRT and ADT depend primarily on the Risk, PSA, and T_stage values.

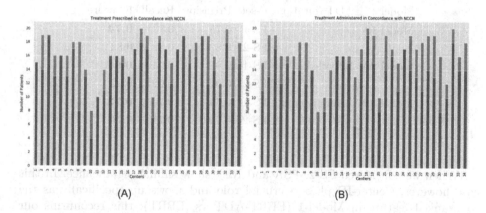

(A)   (B)

Fig. 1. Treatments in concordance with NCCN when all treatments are considered at each center. Blue: treatments in concordance, Orange: not in concordance. (A): Treatments prescribed at each center when ADT intent course is considered along with all other treatments; (B): Treatments administered at each center when ADT administered course is considered along with all other treatments (Color figure online)

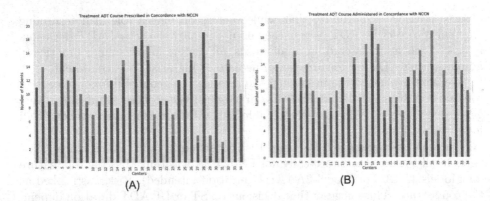

Fig. 2. Patients treated with EBRT and ADT (short term or long term). Blue: number of patients whose treatments are in concordance with NCCN, Orange: number of patients whose treatments are partially not in concordance with NCCN. (A): Treatments prescribed at each center when ADT intent course is considered (B): Treatments administered at each center when ADT administered course is considered (Color figure online)

Table 4. Precision, Recall, F1-Score, for Model-1:(EBRT-ADT vs. EBRT), Model-2: (EBRT-ADT-ST vs. EBRT-ADT-LT) 2A:ADT Intended Duration, 2B:ADT Administered Duration

Model	ADT duration	F-set	Precision	Recall	F1-score
Model 1	–	FS-1	0.75	0.73	**0.74**
		FS-2	0.82	0.82	0.82
Model 2A	Intended	FS-1	0.95	0.94	**0.94**
		FS-2	0.92	0.92	0.92
Model 2B	Administered	FS-1	0.74	0.73	**0.73**
		FS-2	0.72	0.71	0.71

When we considered FS-2, PSA and Risk show similar significance. In this case however, Center-ID plays a crucial role and shows up specifically as the top ranked feature in Model-1 (EBRT-ADT vs. EBRT); this reconfirms our earlier hypothesis that nonclinical factors like the center play a significant role in determining whether patients undergo ADT treatment or not. However, it's significance is much lower in Model-2A (EBRT-ADT-ST vs. EBRT-ADT-LT) with ADT intended duration. Center-ID also shows up as the fourth ranked feature in Model-2B (EBRT-ADT-ST vs. EBRT-ADT-LT) for ADT duration administered; thus we can hypothesize that nonclinical factors may have a role to play in altering the treatment intent.

Table 5. Feature importance in each model. Model 1:EBRT-ADT vs. EBRT, Model 2A: ADT course intended, Model 2B: ADT course Administered

FS	Features	Model 1	Model 2A ADT Intent	Model 2B ADT Administered
FS-1	PSA	**0.52**	0.14	**0.39**
	Risk	0.25	**0.79**	0.30
	Total GS	0.03	0.04	0.14
	T_stage	0.09	0.02	0.07
	Primary GS	0.06	0.01	0.05
	Secondary GS	0.05	0.01	0.05
FS-2	PSA	0.23	0.08	0.24
	Risk	0.28	**0.79**	**0.27**
	Total GS	0.02	0.03	0.19
	T_stage	0.07	0.02	0.05
	Primary GS	0.13	0.02	0.04
	Secondary GS	0.04	0.02	0.04
	Center-ID	**0.29**	0.06	0.17

4   Discussion

In this study, we present an exploratory analysis of localized or locally advanced PCa patients from 34 different VHA treatment centers. We compared the treatments prescribed against the NCCN guideline recommendations and observed that most of the treatment plans (prescribed or administered) matched with the NCCN guidelines. We built machine learning based models to predict the treatment plans for patients and also the likelihood of NCCN concordance of their treatment plans. We observed that PSA and Risk were the top-ranked features in determining the treatment plans for PCa patients.

Center-ID improved the performance of the model that predicts if the selected treatment plan has ADT or not; however, it did not impact the models that predict if the prescribed ADT duration was ST or LT. We also observed some variability in ADT treatments prescribed versus actual ADT treatments administered; the Center-ID, however, had a negligible role to play in such alterations and instead PSA and total Gleason score had significant roles to play in such decisions. We also noticed that the performance status measure had a negative effect on model predictability and hence we dropped it from our feature set. We feel that performance status will be a critical feature in treatment outcome predictions in the future, currently which is outside the scope of this work. Additionally, Risk showed up as the primary feature in predicting ST vs. LT ADT duration. We also observed that the primary reason for treatment plans to be non-concordant with NCCN is due to the ADT course duration not following the guidelines.

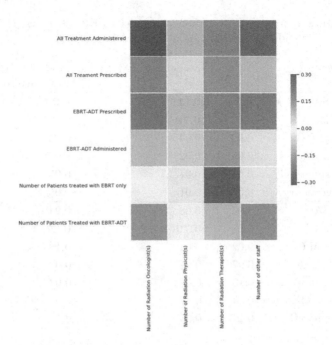

Fig. 3. Pearson correlation between center details (Number of radiation oncologists, radiation physicists, radiation therapists and Other staff) vs. treatment non-concordance (number of non-concordant patients considering all treatments prescribed, all treatments administered, EBRT-ADT prescribed, and EBRT-ADT administered), and treatment selections (number of patients treated with EBRT-only or with EBRT-ADT).

To better understand the impact of non-clinical features like Center-ID in predicting whether the treatment plans were concordant with NCCN guidelines or not, we computed the Pearson correlation between center-specific details (such as staffing details) and the number of non-concordant patients undergoing EBRT-ADT or EBRT-only treatments (either prescribed or administered). Figure 3 shows a small negative correlation between staff details and non-concordance; specifically fewer number of radiation oncologists or radiation therapists led to higher number of non-concordant patients in all cases; while the number of radiation physicists or other staff members did not show any worthwhile correlation. This can be potentially attributed to higher workloads and scheduling conflicts for radiation oncologists/therapists leading to non-adherence to ADT treatment duration requirements from NCCN.

Figure 3 also shows the impact of Center-ID in predicting whether a patient will undergo EBRT-only or EBRT-ADT treatment. We can observe a strong positive correlation between EBRT-only treatment selection and the number of radiation therapists and a less pronounced positive correlation between EBRT-ADT treatment selection and the number of radiation oncologists. While this positive correlation was expected as more radiation oncologists or therapists will lead to more patients being treated with EBRT-ADT or EBRT-only respectively, it is however not clear why the number of radiation physicists or other staff members correlates poorly with these treatment types. It can arise from the bias of the selected patient cohort.

Our findings corroborate previous studies showing the impact of non-clinical factors on prostate cancer treatment patterns. For example, a recent study done on SEERs data reported that prostate cancer treatment patterns were not strictly influenced by outcomes data and varied significantly by patient age, insurance status, financial model, regional bias and socioeconomic factors [11]. An earlier survey on factors influencing treatment selection for localized prostate cancer suggests that recognizing the beliefs that patients hold about their cancer and its treatment could guide the counseling of patients about the treatments available to them and ultimately, help patients make more informed decisions about both their treatments and subsequent adjustments [14]. Prior work on NCCN non-concordance was conducted on elderly patients with high-risk prostate cancer from SEERs was reported that NCCN concordance in elderly patients with aggressive prostate cancer is low [15]. These findings underline the importance of non-clinical factors in treatment decisions, however, reported results were based on single center data; hence they could not identify the center-specific bias as reported in this paper. However, such non-clinical factors can vary appreciably between multiple centers and result in the bias; our future work will include such non-clinical features from the VHA centers to identify the proper reasons behind such center-specific bias.

The VHA ROPA dataset was extracted from recently treated patients having very little to no follow-up data for oncological outcome analysis. Similar predictive models will be built in the future for treatment outcome analysis considering a patient cohort that was treated at earlier dates. Additionally, the ADT duration is generally dependent on the type of drugs used. In this study, we calculated ADT administered duration based on the ADT injection dates; the calculated ADT duration may slightly change considering the ADT injection types. Finally, our study depicts the importance of non-clinical factors, such as Center-ID, in predictive models for treatment selection or concordance to NCCN guidelines. In the future, we will investigate the effects of other types of non-clinical factors (not limited to staffing) pertinent to the specific VHA centers considered here.

References

1. Siegel, R.L.: Cancer statistics. **68**, 7–30 (2018)
2. Tewari, A. (ed.): Prostate Cancer: A Comprehensive Perspective. Springer, Heidelberg (2013). https://doi.org/10.1007/978-1-4471-2864-9
3. Welch, H.G., Albertsen, P.C.: Prostate cancer diagnosis and treatment after the introduction of prostate-specific antigen screening: 1986–2005. J. Natl. Cancer Inst. **101**(19), 1325–1329 (2009)
4. Ross, P.L., et al.: Comparisons of nomograms and urologists' predictions in prostate cancer. In: Seminars in Urologic Oncology, vol. 20, no. 2 (2002)

5. D'Amico, A.V., et al.: Biochemical outcome after radical prostatectomy, external beam radiation therapy, or interstitial radiation therapy for clinically localized prostate cancer. Jama **280**(11), 969–974 (1998)
6. Snow, P.B., Smith, D.S., Catalona, W.J.: Artificial neural networks in the diagnosis and prognosis of prostate cancer: a pilot study. J. Urol. **152**(5), 1923–1926 (1994)
7. Spatz, E.S., Krumholz, H.M., Moulton, B.W.: Prime time for shared decision making. JAMA **317**(13), 1309–1310 (2017)
8. Athas, W.F., et al.: Travel distance to radiation therapy and receipt of radiotherapy following breast-conserving surgery. J. Natl. Cancer Inst. **92**(3), 269–271 (2000)
9. Blumenthal-Barby, J.S., et al.: The neglected topic: presentation of cost information in patient decision AIDS. Med. Decis. Making **35**(4), 412–418 (2015)
10. Altice, C.K., et al.: Financial hardships experienced by cancer survivors: a systematic review. JNCI: J. Natl. Cancer Inst. **109**(2) (2017)
11. Burt, L.M., Shrieve, D.C., Tward, J.D.: Factors influencing prostate cancer patterns of care: an analysis of treatment variation using the SEER database. Adv. Radiat. Oncol. **3**(2), 170–180 (2018)
12. NCCN guidlines for prostate cancer (2016). https://www.cancer.gov/about-cancer/understanding/statistics. Accessed 17 Mar 2018
13. Breiman, L.: anDdom forests. Mach. Learn. **45**(1), 5–32 (2001)
14. Robles, L.A., et al.: Factors influencing patients' treatment selection for localised prostate cancer: a systematic review. Br. J. Med. Surg. Urol. **5**(5), 207–215 (2012)
15. Chen, R.C., Carpenter, W.R., Hendrix, L.H., Wang, Z.A., Nielsen, M.E., Godley, P.A.: Receipt of guideline-concordant treatment in elderly African American and Caucasian patients with prostate cancer. J. Clin. Oncol. **30**(34_suppl), 222 (2012). https://doi.org/10.1200/jco.2012.30.34_suppl.222

Forecasting Model for the Annual Growth of Cryogenic Electron Microscopy Data

Qasem Abu Al-Haija[1] and Kamal Al Nasr[2]

[1] Department of Computer and Information Systems Engineering (CISE), Tennessee State University, Nashville, TN, USA
qabualha@my.tnstate.edu
[2] Department of Computer Science, Tennessee State University, Nashville, TN, USA
kalnasr@tnstate.edu

Abstract. In this paper, we develop a forecasting model for the growth of Cryogenic Electron Microscopy (Cryo-EM) experimental data time series using autoregressive (AR) model. We employ the optimal modeling order that maximizes the estimation accuracy while maintaining the least normalized prediction error. The proposed model has been efficiently used to forecast the growth of cryo-EM data for the next 10 years, 2019–2028. The time series for the number of released three-dimensional Electron Microscopy (3DEM) images along with the time series of the annual number of 3DEM achieving resolution 10 Å or better are used. The data was collected from the public Electron Microscopy Data Bank (EMDB). The simulation results showed that the optimal model orders to estimate both datasets are $AR(5)$ and $AR(6)$ respectively. Consequently, the optimal models obtained an estimation accuracy of 96.8%, and 85% for 3DEM experiments time series and 3DEM resolutions time series, respectively. Hence, the forecasting results reveal an exponential increasing behavior in the future growth of annual released of 3DEM and, similarly, for the annual number of 3DEM achieving resolution 10 Å or better.

Keywords: Protein structure · Electron Microscopy · 3DEM · Single particle · Tomography · X-ray crystallography · NMR · Auto-regressive modeling · Auto-regressive prediction

1 Introduction

Proteins are complex molecules play a vital role and involved in every process within cells. Proteins are linear copolymers built from a sequence of molecules called amino acids, referred to residue, arranged linearly. Proteins are responsible for the vital biological functionalities in every cell. Proteins perform vast type of essential biological functions including signal transporting, providing mechanical support, immune protection, cell adhesion and cell cycle [1, 2]. Proteins in nature fold into unique and energetically favorable three-dimensional (3-D) structures which are crucial and unique to their biological function [3, 4]. The unique conformation in which the protein folds into is called a native structure. The knowledge about the native structure of a given protein is essential

© Springer Nature Switzerland AG 2020
I. Măndoiu et al. (Eds.): ICCABS 2019, LNBI 12029, pp. 147–158, 2020.
https://doi.org/10.1007/978-3-030-46165-2_12

to understand the structure-function relationships. Although the importance of structural information, the ratio between proteins with unknown structures to those with known structure is tremendous. The number of known sequences is about 147 million while the number of resolved structures released by Protein Data Bank is little above 150 K [5–7]. Protein Data Bank (PDB) is a global portal maintain the 3D structural information of biological molecules. It was announced in 1971 and since then it has been used to deposit the newly discovered/determined structures. It contains molecules other than proteins such as nucleic acids and DNA. Traditionally, one of three techniques determine the structure of protein: X-ray crystallography, Nuclear Magnetic Resonance (NMR) and Electron Microscopy (EM). X-ray crystallography is the dominant technique has been used to determine the structure of the proteins. Nearly, 90% of released structures to PDB have been determined by X-ray crystallography. NMR is the second major technique, until recently, and has been used to determine 8% of protein structures. Although X-ray crystallography and NMR have been the dominant techniques for protein structure determination, they endure numerous limitations [8–11]. Some of these limitations are the amount of the sample, the size of the molecule, and crystallization. These limitations become troublesome for macromolecule machines and some protein molecules such as viral capsids, ribosomes, and membrane proteins because most of these do not crystallize easily. In addition, the size of such macromolecules is an existent problem for NMR dynamics.

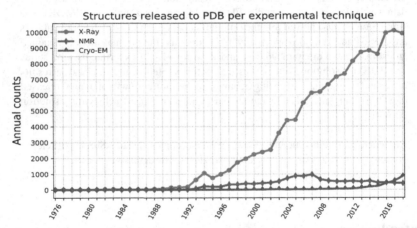

Fig. 1. Annual release rate of molecular structures and their determination techniques. The number of structures released to PDB by the three main techniques annually. X-ray crystallography remains the dominant technique to date. EM becomes more popular than NMR recently. Since 2017, EM is used to produce more structures than NMR.

Recently, Electron Microscopy (EM) is gaining much attention. EM is an experimental imaging technique that aims at visualizing and interpreting unstained macromolecular structures [12]. It produces 3D images (henceforth affectionately referred to as 3DEM) of specimen by averaging thousands of 2D images that are taken from different orientation [13–15]. The resolution of 3DEM determines the information can be visualized/seen about the specimen. The resolution can be obtained by nowadays' technology ranges from near-atomic (<5 Å), sub-nanometer (5 Å–10 Å) and nanometer (>10 Å) resolution [16–18]. Figure 1 shows the annual number of structures released by each of the three main techniques to PDB since 1976 through 2018. To date, PDB contains 150,861 determined structures/entries. As illustrated in Fig. 1, X-ray crystallography (blue line) and NMR (Orange line) have been the main techniques to reveal the high-resolution atomic configuration of protein molecules. EM has recently become the second main technique to determine molecular structure behind X-ray crystallography. Since 2017, it has been used to determine the structure of 1,706 molecules. On the other hand, NMR is used to determine the structure of 918 molecules. Due to current technologies and improved detectors, these numbers are expected to increase and the gap to grow.

EM can capture the coexisting structural states of biological machines [25, 26]. Accordingly, it is proven to be a major source of the information about functional mechanisms and motions of these machines. Although not all of images produced by EM have been resolved at near-atomic (< 5 Å) resolution [27, 28], recent advances in EM will lead to more images produced at near-atomic and sub-nanometer resolution. The number of quality images to be analyzed rapidly grows with the greatly improved detectors that are now available [18, 29]. In addition, the significant increase in signal-to-noise ratio (SNR) represents an important contribution to EM [30–32]. It has been reported recently that the resolution of structures determined by EM is approaching those determined by X-ray crystallography [17]. Published 3DEM entries have resolutions approaching from near-atomic to sub-angstrom resolution range for appropriate samples and can be directly used to produce protein models. On the other hand, at resolutions worse than near-atomic (> 5 Å), substantial help can be provided by the computational algorithms and the hardware power we have nowadays [33–36]. This will augment the number of available structural models. As shown in Fig. 4, the annual number of 3DEM in EMDB achieving resolution 10 Å or better is increasing. The improvement of the resolution is obvious for single particle technique.

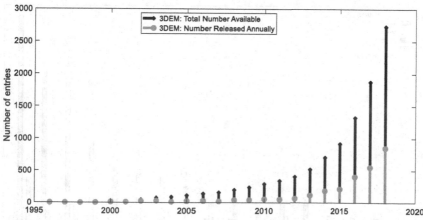

Fig. 2. Overall growth rate of 3DEM released. The cumulative number of 3DEM released to EMDB over years, since 1996, is shown as a black bar-chart with filled diamonds. In addition, the annual deposition growth is shown as a green bar-chart with filled circles. (Color figure online)

EM has been started as early as late 1990's. Since the first 3DEM reported for hepatitis B virus [19, 20], it was used to study many of macromolecular complexes [21–24]. The Electron Microscopy Data Bank (EMDB) is the unified global portal for deposition and retrieval of 3DEM, atomic models, and associated metadata [21]. The EMDB currently holds more than 7,000 3DEM entries in addition to more than 3,000 PDB entries of fitted molecular models (e.g., atomic configurations). Figure 2 depicts the release rate of 3DEM in EMDB through years 2002 to 2018. EM consists of four (4) main types: Transmission Electron Microscope (TEM), Scanning Electron Microscope (SEM), Reflection Electron Microscope (REM) and Scanning Transmission Electron Microscope (STEM). These types differ slightly in the way the specimen is prepared and the way the 3DEM is produced. Many techniques have been used to visualize molecules and generate 3DEM such as tomography or single particle. The most technique has been used to visualize macromolecules is single particle. Single particle is a TEM technique. Figure 3 shows the major techniques have been used and their annual number of 3DEM released. As illustrated in the figure, single particle has been the most technique used since 2002.

Undoubtedly, the determination of protein structure is considered as one of the most important objectives tracked by bioinformatician, as it is highly important in medicine (i.e., drug design) and biotechnology (i.e., the design of novel enzymes) [37]. Therefore, it is important to study the growth trend of high resolution 3DEM in EMDB. Generally, deposition rate of 3DEM can be analyzed as time series [38] with continuous evolution of entries along with time. Thus, the 3DEM series can be modeled and predicted for further analysis of the growth behavior. However, the trend prediction can be accomplished using different efficient signal modeling and processing techniques such as the parametric techniques of auto-regressive (AR) modeling. Auto-regressive (AR) model has been widely used for signal modeling and prediction for many different phenomena; including engineering, social, environmental, and financial cases. Examples of using AR model

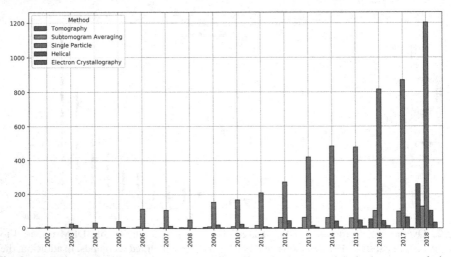

Fig. 3. Experimental techniques used to produce 3DEM. Single particle is the most technique used to generate 3DEM. Tomography is the second most popular technique.

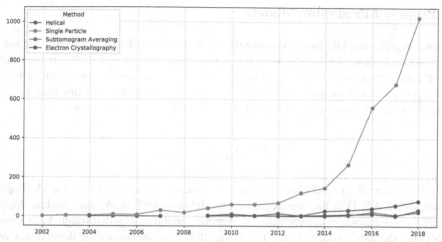

Fig. 4. Experimental methods used to produce 3DEM and improvement in resolution. The number of 3DEM resolved at (0 Å–10 Å) resolution is shown for different methods. The number of 3DEM is increasing for all methods. Single particle has the most improvement growth rate. Some methods have not released/resolved 3DEM entries during years (i.e., 2007–2009), therefore, some lines show gaps.

for time series prediction purposes can be found in [39–44]. One objective of this paper is to show that the growth rate of cryo-EM technology is exponential. This is important for EM community to have statistical numbers and to know the exact trend of the growth of EM data. We are using the Auto-regressive (AR) model to regenerate and analyze the time series of annual number of 3DEM released for the last 18 years (i.e., from 2002 to 2018) and the annual number of 3DEM achieving 10 Å resolution or better for the same time frame using optimal modeling order that maximizes the model fitting percentages while minimizing the final estimation error. Specifically, the main contributions of this paper can be summarized as follows (Fig. 4):

- We develop an AR model for the collected 3DEM time series that can maintain optimal degree of AR modeling with minimum modeling error to optimize the time series estimation and forecasting.
- We employ the optimal derived AR model to re-generate the time series of the measured data and predict the short-term future values up to year 2028 (i.e., ten years ahead) of possible numbers of the following data-sets: the annual growth of 3DEM released for the last 18 years (i.e., from 2002 to 2018) and the annual number of 3DEM achieving 10 Å resolution or better (i.e., from 2002 to 2018).
- We provide simulation plots for the original collected signal along with the forecasted signal with analysis to gain insight into the developed model and the solution technique.

The rest of this paper is organized as follows: Sect. 2 presents and discusses the system model using AR technique and the simulation results by considering several scenarios. Finally, Sect. 3 concludes the paper.

2 Proposed Prediction Models

The autoregressive (AR) model is a recursive parametric modeling technique that uses the feedback to generate the internal dynamics for the predicted signal [39]. Indeed, AR model is a representation of a type of random process. It specifies that the output variable depend linearly on its own previous values and on a stochastic term. The notation $AR(p)$ indicates an autoregressive model of order p. The $AR(p)$ model is defined as:

$$X_n = c + \sum_{i=1}^{p} a_i X_{n-i} + e_n$$

Where a_i are the model parameters, c is a constant, and e_n is white noise. A simple model, called $AR(1)$ or the "autoregressive model of order 1," can be derived as: $X_n = c + a_1 X_{n-1} + e_n$, where current value X_n depends on the past only through X_{n-1} since $AR(1)$ use the model order of one. The higher the order, the higher the dependency on the past data. However, the appropriate model order for AR modeling differs among time series based on the number of observations in the time series and the level of linearity among them [44].

In this work, we have developed an AR model to re-generate the time series signals for the annual number of 3DEM released from 2002 to 2018 and the annual number of 3DEM achieving 10 Å resolution or better from 2002 to 2018 using optimal modeling order, and thus predicting the next decade of signal tendency. In the previous section, we showed the actual datasets for both time-series (i.e., the annual growth of 3DEM released and the annual number of 3DEM with resolution better than 10 Å from 2002 to 2018). For optimal time series modeling and estimation, we need to employ the optimal order of AR model. Therefore, we have generated and plot the AR model estimation errors vs. model orders in order to select the optimal model order number that minimizes the error and the design cost as well. The prediction error can be calculated by different methods such as by calculating $(Norm(e)/Norm(y))$ for each model order.

Figure 5 shows the relationship between the different order model values against the final prediction error. The optimum model order is the order in which the model error has the smallest value with acceptable design cost (i.e., when the error of two model orders are close, then its preferable to select the lowest order that maximizes the accuracy and minimizes the design cost). In our developed model, we have selected $AR(5)$ and $AR(6)$ as the optimal model orders to re-generate and estimate the annual numbers of 3D images of Electron Microscopy (3DEM) time series and the annual number of 3DEM data achieving resolution ≤ 10 Å time series respectively since, they correspond to the least normalized final perdition errors (FPE) [45, 46]. Both of models have recorded an accurate curve fitting percentage of 96.8% and 85% for 3DEM dataset and 3DEM with resolution ≤ 10 Å time series, respectively. Such results conform to the expected increasing growth in the tendency of both released numbers of EM Data. It should be noted that the reason of achieving higher curve fitting accuracy for 3DEM dataset from 3DEM with resolution ≤ 10 Å time series 96.8% vs. 85%), is due to the higher level of linear correlation data items for 3DEM dataset.

Figure 5 shows the relationship between the different order model values against the final prediction error. The optimum model order is the order in which the model error

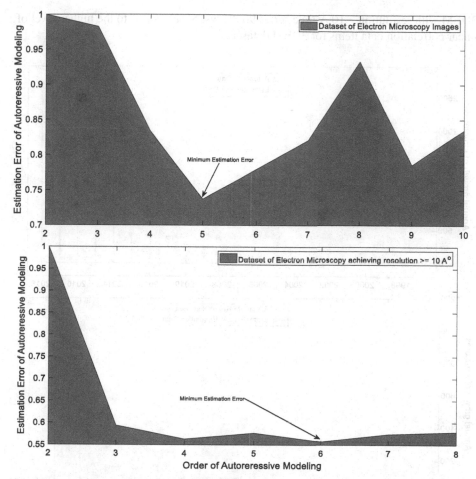

Fig. 5. Normalized Final Prediction Error (FBE) vs. Model Order Number. We have selected the optimal estimation order based on the relationship between model order and prediction error. Thus, the model number of 3DEM time series and the annual number of 3DEM with resolution ≤ 10 Å have been estimated with optimal model orders: $AR(5)$ and $AR(6)$ respectively.

has the smallest value with acceptable design cost (i.e., when the error of two model orders are close, then its preferable to select the lowest order that maximizes the accuracy and minimizes the design cost). In our developed model, we have selected $AR(5)$ and $AR(6)$ as the optimal model orders to re-generate and estimate the annual numbers of 3D images of Electron Microscopy (3DEM) time series and the annual number of 3DEM data achieving resolution ≤ 10 Å time series respectively since, they correspond to the least normalized final perdition errors (FPE) [45]. Both of models have recorded an accurate curve fitting percentage of 96.8% and 85% for 3DEM dataset and 3DEM with resolution ≤ 10 Å time series, respectively. Such results conform to the expected increasing growth in the tendency of both released numbers of EM Data. It should be noted that the reason of achieving higher curve fitting accuracy for 3DEM dataset from

3DEM with resolution ≤10 Å time series 96.8% vs. 85%), is due to the higher level of linear correlation data items for 3DEM dataset.

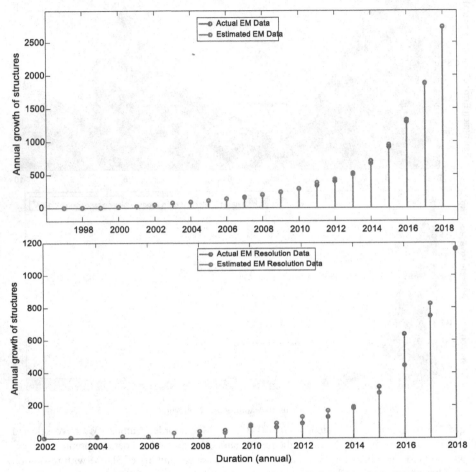

Fig. 6. Modeling the datasets: Actual Data (blue) vs. Estimated Data (red). In this figure, each dataset was regenerated using its corresponding optimal order to see how modeled signal is tightly coupled with its pear of original measured data. Accordingly, all figures of the datasets proved a high-level of confidence to estimate since they show very similar tendency to their peers of the original datasets. (Color figure online)

Figure 6 illustrates the plots for both time series signals in terms of the actual measured dataset and the model estimated data using $AR(5)$ and $AR(6)$ models for the global number of 3DEM time series and the annual number of 3DEM achieving resolution ≤10 Å time series respectively. As depicted from the figure, the estimation for both time series signals are precise and consistent for almost all the measured dataset signals with minor variations observed to appear in the modeling of time series plot for the annual number of 3DEM achieving resolution ≤10 Å. The reason of having such a significant signal compatibility is due to the large level of signals' linearity for the original measured data which impact the estimation process of the linear AR parametric model especially for the time series of the number of 3DEM released annually, and thus, can be used to forecast the numbers for the next decade with high level of confidence (i.e., accuracy) peaked at 96.8% of model fitting (Fig. 7).

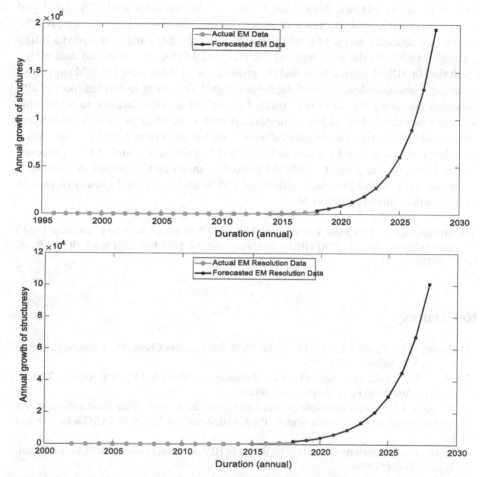

Fig. 7. Forecasting the future decade for both-time series (up to year 2028). According to the figures, the growth rate of number of entries released for 3DEM and for the annual number of 3DEM achieving 10 Å resolution or better showed an exponential increasing trend.

3 Conclusion

An optimal auto-regressive (AR) process to model and forecast of two Cryo-EM data series was proposed in this paper. In this paper, we proposed a predictive model of the time series of the number of entries of 3DEM released annually to EMDB and the number of 3DEM achieving 10 Å resolution or better using auto-regressive (AR) model. We employed the optimal modeling order that maximizes the estimation accuracy while maintaining minimum prediction error. The proposed model was developed to forecast the future 10 years, 2019–2028, for the two datasets in the study. The developed scheme recorded a fitting accuracy of 96.8% for the fifth order of AR process, AR(5), for the first dataset, number of released 3DEM. On the other hand, it recorded a fitting accuracy of 85% for the sixth order of AR process, AR(6), for the second dataset, the resolution of 3DEM entries. Note that the growth rate of the released 3DEM suggests a faster exponential tendency. Therefore, the developed forecasting model has successfully predicted the future decade of the expected growth of the two datasets. In future, this work can be extended by adding a comparison with some other prediction models and state-of-art works. Also, more statistical results, such a mean value for all simulation and confidence intervals, can be added and analyzed to gain more insight of the proposed model. Final, we are going to study the trend for some other important datasets such as the annual released protein structures into PDB and the annual growth of protein structures determined by cryo-EM.

Acknowledgments. This work was supported by the US National Science Foundation (NSF) Research Initiation Award (RIA) (HRD: 1600919) and the NIH Research grant (R15-AREA: 1R15GM126509-01).

References

1. Rappe, A.K., Casewit, C.J.R.: Molecular Mechanics Across Chemistry. University Science Books, Mill Valley (1997)
2. Siegel, G.J., et al.: Basic Neurochemistry: Molecular, Cellular and Medical Aspects, 7th edn. Elsevier Academic Press, Amsterdam (2006)
3. Blundell, T.L., et al.: Insulin-like growth factor: a model for tertiary structure accounting for immunoreactivity and receptor binding. Proc. Natl. Acad. Sci. U.S.A. (PNAS) **75**(1), 180–184 (1978)
4. Weber, I.T.: Evaluation of homology modeling of HIV protease. Proteins Struct. Funct. Bioinf. **7**(2), 172–184 (1990)
5. Sussman, J.L., et al.: Protein data bank (PDB): database of three-dimensional structural information of biological macromolecules. Acta Crystallogr. Sect. D: Biol. Crystallogr. **54**(6–1), 1078–1084 (1998)
6. Berman, H., et al.: The protein data bank. Nucleic Acids Res. **28**, 235–242 (2000)
7. Bernstein, F.C., et al.: The Protein data bank: a computer-based archival file for macromolecular structures. J. Mol. Biol. **112**(3), 535–542 (1977)
8. Zheng, H., et al.: X-ray crystallography over the past decade for novel drug discovery – where are we heading next? Expert Opin. Drug Discov. **10**(9), 975–989 (2015)

9. Pearson, A.R., Mozzarelli, A.: X-ray crystallography marries spectroscopy to unveil structure and function of biological macromolecules. Biochimica et Biophysica Acta (BBA) - Proteins Proteomics **1814**(6), 731–733 (2011)
10. Emwas, A.-H.M.: The strengths and weaknesses of NMR spectroscopy and mass spectrometry with particular focus on metabolomics research. In: Bjerrum, J.T. (ed.) Metabonomics. MMB, vol. 1277, pp. 161–193. Springer, New York (2015). https://doi.org/10.1007/978-1-4939-2377-9_13
11. Yusupova, G., Yusupov, M.: Ribosome biochemistry in crystal structure determination. RNA (New York, N.Y.) **21**(4), 771–773 (2015)
12. Wang, L., Sigworth, F.J.: Cryo-EM and single particles. Physiology **21**(1), 13–18 (2006)
13. Mitra, K., Frank, J.: Ribosome dynamics: insights from atomic structure modeling into cryo-electron microscopy maps. Annu. Rev. Biophys. Biomol. Struct. **35**, 299–317 (2006)
14. Khatter, H., et al.: Structure of the human 80S ribosome. Nature **520**, 640 (2015)
15. Liu, Z., et al.: 2.9 Å resolution cryo-EM 3D reconstruction of close-packed virus particles. Structure (London, England: 1993) **24**(2), 319–328 (2016)
16. Kühlbrandt, W.: Cryo-EM enters a new era. eLife **3**, e03678–e03678 (2014)
17. Liu, S., et al.: Atomic resolution structure determination by the cryo-EM method MicroED. Protein Sci. **26**(1), 8–15 (2017)
18. Kühlbrandt, W.: The resolution revolution. Science **343**(6178), 1443 (2014)
19. Bottcher, B., Wynne, S.A., Crowther, R.A.: Determination of the fold of the core protein of hepatitis B virus by electron cryomicroscopy. Nature **386**(6620), 88–91 (1997)
20. Conway, J.F., et al.: Visualization of a 4-helix bundle in the hepatitis B virus capsid by cryo-electron microscopy. Nature **386**(6620), 91–94 (1997)
21. Lawson, C.L., et al.: EMDataBank.org: unified data resource for CryoEM. Nucleic Acids Res. **39**(suppl 1), D456–D464 (2011)
22. Zhang, X., et al.: 3.3 Å cryo-EM structure of a nonenveloped virus reveals a priming mechanism for cell entry. Cell **141**(3), 472–482 (2010)
23. Baker, M.L., et al.: Ab initio modeling of the herpesvirus VP26 core domain assessed by CryoEM density. PLoS Comput. Biol. **2**(10), e146 (2006)
24. Villa, E., et al.: Ribosome-induced changes in elongation factor Tu conformation control GTP hydrolysis. Proc. Natl. Acad. Sci. (PNAS) **106**(4), 1063–1068 (2009)
25. Amunts, A., et al.: Structure of the yeast mitochondrial large ribosomal subunit. Science **343**(6178), 1485–1489 (2014)
26. Bell, D.C., et al.: Successful application of low voltage electron microscopy to practical materials problems. Ultramicroscopy **145**, 56–65 (2014)
27. Fischer, N., et al.: Structure of the E. coli ribosome-EF-Tu complex at <3 Å resolution by Cs-corrected cryo-EM. Nature **520**(7548), 567–570 (2015)
28. Bartesaghi, A., et al.: 2.2 Å resolution cryo-EM structure of β-galactosidase in complex with a cell-permeant inhibitor. Science **348**(6239), 1147–1151 (2015)
29. Callaway, E.: The revolution will not be crystallized: a new method sweeps through structural biology. Nature **525**(7568), 172–174 (2015)
30. Bammes, B.E., et al.: Direct electron detection yields cryo-EM reconstructions at resolutions beyond 3/4 Nyquist frequency. J. Struct. Biol. **177**(3), 589–601 (2012)
31. Milazzo, A.-C., et al.: Initial evaluation of a direct detection device detector for single particle cryo-electron microscopy. J. Struct. Biol. **176**(3), 404–408 (2011)
32. Bai, X.-C., et al.: Ribosome structures to ar-atomic resolution from thirty thousand cryo-EM particles, vol. 2 (2013). Editor W. Kühlbrandt
33. Al Nasr, K., et al.: Analytical approaches to improve accuracy in solving the protein topology problem. Molecules **23**(2), 28 (2018)
34. Al Nasr, K., et al.: PEM-fitter: a coarse-grained method to validate protein candidate models. J. Comput. Biol. **25**, 21–32 (2018)

35. Biswas, A., et al.: An effective computational method incorporating multiple secondary structure predictions in topology determination for cryo-EM images. IEEE/ACM Trans. Comput. Biol. Bioinf. **14**(3), 578–586 (2017)
36. Al Nasr, K., He, J.: Constrained cyclic coordinate descent for cryo-EM images at medium resolutions: beyond the protein loop closure problem. Robotica **34**(8), 1777–1790 (2016)
37. Pirovano, W., Heringa, J.: Protein secondary structure prediction. Methods Mol. Biol. **609**, 327–348 (2010). ISBN 978-1-60327-240-7. PMID 2019
38. Imdadullah, Time Series Analysis. Basic Statistics and Data Analysis (2014). http://itfeature. com/time-series-analysis-and-forecasting/time-series-analysis-forecasting
39. Al-Haija, Q.A., Mao, Q., Al Nasr, K.: Forecasting the number of monthly active Facebook and Twitter worldwide users using ARMA model. J. Comput. Sci. **15**(4), 499–510 (2019). https://doi.org/10.3844/jcssp.2019.499.510
40. Huang, J., et al.: Forecasting solar radiation on an hourly time scale using a Coupled Autoregressive and Dynamical System (CARDS) model. Sol. Energy **87**, 136–149 (2013). https://doi.org/10.1016/j.solener.2012.10.012
41. Lydia, M., et al.: Linear and non-linear autoregressive models for short-term wind speed forecasting. Energy Convers. Manage. **112**, 115–124 (2016). https://doi.org/10.1016/j.enconman.2016.01.007
42. Abadi, A., et al.: Traffic flow prediction for road transportation networks with limited traffic data. IEEE Trans. Intell. Transp. Syst. **16**(2), 653–662 (2015)
43. Ruiz, L.G.B., et al.: An application of non-linear autoregressive neural networks to predict energy consumption in public buildings. Energies **9**(9), 684 (2016). https://doi.org/10.3390/en9090684
44. Al-Haija, Q.A., Tawalbeh, L.: Autoregressive modeling and prediction of annual worldwide cybercrimes for cloud environments. In: IEEE 10th International Conference on Information and Communication Systems (ICICS 2019) (2019)
45. Niedwiecki, M., Cioek, M.: Akaike's final prediction error criterion revisited. In: 40th International Conference on Telecommunications & Signal Processing (ICTSP 2017) (2017)
46. Al Nasr, K., Al-Haija, Q.A.: Forecasting the growth of structures from NMR and X-ray crystallography experiments released per year. J. Inf. Knowl. Manag. (JIKM) **19**(1), 1–12 (2019). Special Issue

Local and Global Stratification Analysis in Whole Genome Sequencing (WGS) Studies Using LocStra

Georg Hahn[1]([✉]), Sharon Marie Lutz[1], Julian Hecker[2], Dmitry Prokopenko[3],
and Christoph Lange[1]

[1] Department of Biostatistics, T.H. Chan School of Public Health,
Harvard University, Boston, MA 02115, USA
{ghahn,smlutz,clange}@hsph.harvard.edu
[2] Department of Medicine, Channing Laboratory, Brigham and Women's Hospital,
Boston, MA 02115, USA
jhecker@hsph.harvard.edu
[3] Department of Neurology, Massachusetts General Hospital and Harvard Medical
School, Boston, MA 02115, USA
dprokopenko@mgh.harvard.edu

Abstract. We are interested in the analysis of local and global population stratification in WGS studies. We present a new R package (*locStra*)
that utilizes the covariance matrix, the genomic relationship matrix, and
the unweighted/weighted genetic Jaccard similarity matrix in order to
assess population substructure. The package allows one to use a tailored
sliding window approach, for instance using user-defined window sizes
and metrics, in order to compare local and global similarity matrices.
A technique to select the window size is proposed. Population stratification with *locStra* is efficient due to its C++ implementation which
fully exploits sparse matrix algebra. The runtime for the genome-wide
computation of all local similarity matrices does typically not exceed one
hour for realistic study sizes. This makes an unprecedented investigation
of local stratification across the entire genome possible. We apply our
package to the 1,000 Genomes Project.

Keywords: Population stratification · WGS studies · Covariance
matrix · Genomic relationship matrix · genetic jaccard similarity
matrix

1 Introduction

Though being vulnerable to confounding due to population substructure [4],
genetic association studies are a popular mapping tool in population-based
designs. To address confounding, a variety of methods have been proposed [3,10].
Numerous methods that rely on the genetic relationship matrix [17], estimated
from observed genotypes, have been proposed: EIGENSTRAT, STRATSCORE,

© Springer Nature Switzerland AG 2020
I. Măndoiu et al. (Eds.): ICCABS 2019, LNBI 12029, pp. 159–170, 2020.
https://doi.org/10.1007/978-3-030-46165-2_13

multi-dimensional scaling, etc. [5,7,8]. Despite their popularity, current software implementations are computationally intensive and rely on prior linkage disequilibrium (LD) pruning, which reduces the dataset dimension. Running times ranging from hours to days for datasets of typical sizes are commonplace.

Though previous research showed strong evidence for local stratification [6,9,18], the GRM computation is typically done globally to capture global population structure. This is due to the fact that for a comprehensive, genome-wide analysis of local stratification, the computational burden has been too prohibitive. Moreover, often genomic regions do not contain a sufficient number of loci that are not in linkage disequilibrium (LD), thus causing problems for matrix-based approaches which are designed for uncorrelated common variant data.

Densely spaced rare variants (RVs) that are mostly not in LD became commonplace with the arrival of whole genome sequencing data (WGS). RVs are more informative about recent admixture as they are genetically younger than common variants, and approaches based on Jaccard similarity matrices [11,13] that utilize RV/WGS data have been developed. Nevertheless, those approaches continue to suffer from computational inefficiencies.

Our R package *locStra* allows the user to assess population stratification in RVs at the local and global level using four approaches: (1) the covariance matrix, (2) the genomic relationship matrix, (3) the unweighted and (4) weighted Jaccard similarity matrices. All implementations in *locStra* are fully written in C++, and all computations on similarity matrices are fully carried out on sparse data structures to maximize computational efficiency. The generic sliding window algorithm of *locStra* enables the fast analysis of local stratification at the genome-wide level. To select appropriate window sizes we provide a data-driven algorithm.

We illustrate the importance of investigating local substructure with *locStra* using the 1,000 Genomes Project. Moreover, we evaluate the four similarity matrix approaches with regards to their performance and assess *locStra* in terms of runtime. The fast implementations in *locStra* allow for novel research in local stratification in WGS studies and subsequent insights into association findings.

This article is structured as follows. Section 2 presents the functionality of the *locStra* package. In Sect. 3, we illustrate the practical relevance of our contribution through an analysis of the 1,000 Genomes Project dataset. We conclude with a discussion in Sect. 4. The appendix gives further details regarding the sparse matrix computations in our implementation (Sect. A), theoretical runtimes for all similarity matrix approaches (Sect. B), and it contains a comparison of our locStra package with the standard tool *PLINK2* [2,12] in Sect. C.

2   Implementation

Our package *locStra* is available online and can be downloaded from *The Comprehensive R Archive Network*. Its core relies on *RcppEigen* of [1] offering fully sparse matrix algebra in C++.

This section presents the seven functions implemented in *locStra*.

2.1 Dense and Sparse Matrix Implementation

The following functions provide C++ implementations of the four standard approaches mentioned in Sect. 1. We provide separate code for dense and sparse input matrices in order to maximize computational efficiency. The C++ implementation to be used can be selected with the boolean argument *dense*. The default is *dense=False*.

1. The covariance matrix is computed with the function *covMatrix*. The input is allowed to be any real valued matrix.
2. The genomic relationship matrix (GRM), defined in [17], is computed with *grMatrix*. The input must be a binary matrix. The boolean flag *robust* can be used to select either the classic or the robust version of the GRM (the default is *robust=True*).
3. The Jaccard similarity matrix is computed with *jaccardMatrix*. The input must be a binary matrix.
4. The weighted Jaccard matrix [14] is implemented in *sMatrix*. If the input data is phased, this can be indicated with the boolean argument *phased* in the function *sMatrix* (the default is *phased=False*). If a cutoff value for the minimal number of variants to consider is desired, this can be specified with the integer parameter *minVariants* (the default is *minVariants=5*).

2.2 Main Function

Our package offers a flexible way to tailor the local population stratification scan of the data through the generic structure of our main function *fullscan*. It has five arguments:

- The (sparse) matrix containing the sequencing data is the first input. We assume that the input matrix is oriented in such a way as to contain the data for one individual per column.
- The precise windows to be used in the sliding window scan are given in the second argument as a two-column matrix (called *windows*). In the matrix *windows*, the start and end positions of each window are the two entries per row. We provide an auxiliary function called *makeWindows* (Sect. 2.3) to facilitate the generation of the matrix containing the sliding window entries.
- The precise mechanism for processing each sliding window is specified with the function *matrixFunction* (third argument), which operates on one matrix input argument. In principle, any function can be used, though typical choices include *covMatrix*, *grMatrix*, *jaccardMatrix*, or *sMatrix*.
- After processing each window, we need to specify a *summaryFunction* on one input argument that is compatible with the output of *matrixFunction*. In principle, this can again be any function though the computation of the largest eigenvector with the help of the function *powerMethod* (Sect. 2.3) is an intuitive choice.

- Last, the fifth input argument to be specified is the function *comparisonFunction* which is needed to compare the summaries (e.g., first eigenvectors) on a local and a global level. The *comparisonFunction* can be any function on two arguments compatible with the output of the function *summaryFunction*. The native *R* correlation function *cor* for two input vectors is a typical choice.

The global and local comparison values for each window are returned by *fullscan* as a two column matrix. Each row in the matrix *windows* corresponds to a row in the output of *fullscan* in the same order.

2.3 Auxiliary Functions

We provide two auxiliary functions which were already referred to in the description of *fullscan*:

1. The two-column matrix of non-overlapping or overlapping windows can be created with the auxiliary function *makeWindows*. This function has three arguments: the length of the data, the window size, and an offset. It creates non-overlapping windows if the offset is equal or larger than the window size. Otherwise, sliding windows of given window size and offset are created.
2. The power method is a fast iterative algorithm to compute the largest eigenvector [16]. Our *R* package provides a C++ implementation in the auxiliary function *powerMethod*.

3 Local Stratification Analysis of the 1,000 Genome Project

In this section we highlight the practical relevance of *locStra* by applying it to all chromosomes of the 1,000 Genome Project [15]. We take a closer look at the results for four chromosomes (precisely, chromosomes 5, 10, 12, and 16) to stress the feasibility of local substructure analysis at the genome-wide level (Sect. 3.1). Section 3.2 reports runtime results across all chromosomes, and Sect. 3.3 illustrates a procedure to select suitable window sizes for population stratification.

We prepare the raw data from the 1,000 Genome Project with PLINK2. In particular, we selected rare variants with a cutoff value of 0.01 for option *--max-maf*. LD pruning was applied with parameters *--indep-pairwise 2000 10 0.01*. In the following we report results for the EUR super population of the 1,000 Genome Project.

3.1 Data Analysis Results for Certain Chromosomes of the 1,000 Genome Project

We apply our sliding window approach with a window size of 120,000 RVs. This value is suggested by the window selection algorithm we propose in Sect. 3.3. Figure 1 displays the correlations between the first eigenvectors of all local similarity matrices with the corresponding first eigenvector of the global similarity

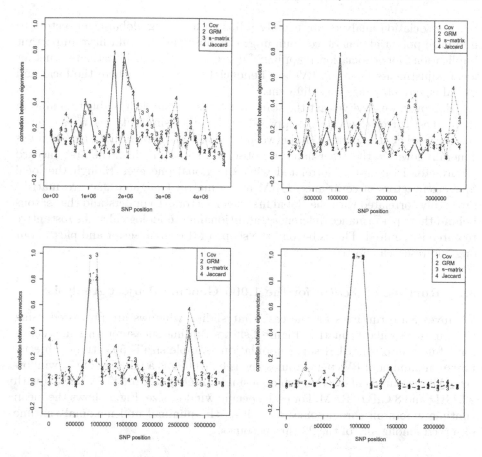

Fig. 1. Correlation of local to global eigenvector of the covariance matrix, GRM matrix, s-matrix, and Jaccard matrix. Chromosomes 5 (top left), 10 (top right), 12 (bottom left), and 16 (bottom right). Super population EUR of the 1,000 Genomes Project. Window size 128000 RVs.

matrix for the four different types of similarity matrices and the four selected chromosomes.

We observe several noteworthy features in the local substructure analysis. The local substructure analysis shows only a couple of genomic regions in which the local and global substructures are similar when measured with the help of the first eigenvectors, independently of the similarity matrix used. When measured via similarity matrices, we observe considerable variability overall in the local substructure across the genome.

For chromosome 16 (Fig. 1, bottom right), we observe a small genomic region where all correlations reach nearly 1. Otherwise, the correlations between the first local and first global eigenvectors are very small throughout the genome for all four similarity matrices. This could hint at the fact that the global substructure is very different from the local substructure as captured by similarity matrices.

Since association analyses are usually adjusted for using global eigenvectors to minimize potential genetic confounding, this observation could have important implications for association mapping. Future research will investigate whether local adjustments based on RVs are beneficial in scenarios where the local and global eigenvectors are very different.

Moreover, we observe that the standard Jaccard approach is able to maintain the highest correlation values compared to the other similarity matrices even in the areas where the correlation between the local and global first eigenvector is generally low. On the other hand, we observe regions in which the first Jaccard eigenvector is almost uncorrelated with the global one even though the local first eigenvectors of covariance, GRM and s-matrix (weighted Jaccard matrix) are highly correlated with the global first eigenvectors. To understand the reasons behind these performance differences, additional methodological and substantive research is required. This is beyond the scope of this manuscript and part of our ongoing research efforts.

3.2   Runtime of *locStra* for the 1,000 Genome Project Analysis

We investigate runtimes for the case that sliding windows are processed using the Jaccard similarity matrix. Figure 2 shows runtime measurements in seconds for a full scan of the EUR super population (sample size is 503 study subjects; maximal number of RVs per chromosome is 6267065) as a function of the window size. All measurements were taken on a single Intel QuadCore i5-7200 CPU with 2.5 GHz and 8 GiB of RAM. For each specific window size, Fig. 2 shows the mean runtime across all chromosomes as well as the minimal and maximal runtime observed among any of the 22 chromosomes.

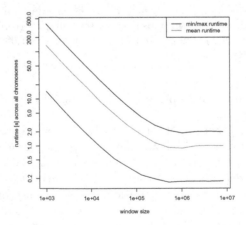

Fig. 2. Runtime for computing the Jaccard matrix across all chromosomes as a function of the window sizes. Minimal, maximal, and mean runtimes across all chromosomes. Super population EUR of the 1,000 Genomes Project. Logarithmic scale on the x- and y-axes.

We observe that the runtime decreases for larger window sizes as expected. Moreover, a complete scan of any chromosome has a mean runtime not exceeding 500 s. The runtime of any method is in the vicinity of one minute for a full scan when using the realistic window size of 10^5 RVs suggested in Sect. 3.3.

For the other three approaches (covariance matrix, genomic relationship matrix, weighted Jaccard matrix) the runtimes are similar and thus not reported. Likewise, runtime measurements for the AFR super population of the 1,000 Genome Project show qualitative similar results which are not reported.

3.3 Selecting Suitable Window Sizes for Population Stratification

We address the question of selecting an appropriate window size for the population stratification scan. To this end, we observe that as the window size increases, less windows are used in the scan of the data, thus causing the correlation between local and global eigenvectors to increase (Fig. 3, left). On the contrary, we obtain less data points when using fewer windows across the genome (as implied by the larger window sizes), thus making results less meaningful. The two quantities, increasing correlation and decreasing number of datapoints when using larger window sizes, thus work against each other.

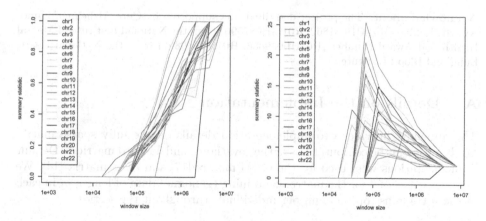

Fig. 3. Mean correlation across all windows per chromosome (left) and mean correlation across all windows multiplied by the number of windows (right) as a function of the window size. Super population EUR of the 1,000 Genomes Project. Correlation values computed between global and local eigenvectors of the Jaccard matrices.

In such cases, for a particular window size, the product of the mean correlation among all chromosomes and the number of generated windows is a natural tradeoff which we use as a measure for decision making (Fig. 3, right). The figure shows that the product of mean correlation and number of windows is close to zero for small and large window sizes. However, we observe a peak at a window size of around 10^5 RVs for the 1,000 Genomes Project which seems to appear at an almost identical position across all chromosomes. The reason for the almost identical position is the fact that the slope in Fig. 3 (left) is very similar across all chromosomes, thus causing the product of mean correlation and number of windows to have a similar functional shape in Fig. 3 (right).

Based on our proposed measure, we recommend a window size of around 10^5 RVs for the analysis of the 1,000 Genomes Project data. Qualitatively similar results (not reported here) are obtained when carrying out the same analysis for the AFR super population of the 1,000 Genome Project.

We suggest the heuristic approach of Fig. 3 (right) as a general technique for selecting a window size in population stratification with sliding windows.

4   Conclusion

A comprehensive genome-wide analysis of local stratification in WGS studies based on similarity matrices is possible using our R package *locStra* due to its runtimes of around 500 s for the genome-wide analysis of all sliding windows in the EUR super population (one Intel QuadCore i5-7200 CPU with 2.5 GHz and 8 GiB of RAM). This will enable the community to investigate local stratification patterns at a genome-wide level in WGS studies.

Acknowledgment. The project described was supported by Cure Alzheimer's fund, Award Number (R01MH081862, R01MH087590) from the National Institute of Mental Health and Award Number (R01HL089856, R01HL089897) from the National Heart, Lung and Blood Institute.

A   Details on the Implementation

The appendix provides two implementation details on the fully sparse matrix algebra used in the computations of the covariance and Jaccard matrices. Default implementations were used for the GRM matrix [17] and the s-matrix [14]. We assume $X \in \mathbb{R}^{m \times n}$ for the matrix containing (genomic) data of length m in each of the n columns (one column per individual) throughout the section.

A.1   Covariance Matrix

We first look at computing the covariance matrix in dense algebra. Let the column means of X be given as vector $v \in \mathbb{R}^n$ and denote as $Y \in \mathbb{R}^{m \times n}$ the matrix consisting of the rows of X with their mean substracted. Then

$$\text{cov}(X) = \frac{1}{m-1} Y^\top Y.$$

In the sparse case, it is not possible to compute $\text{cov}(X)$ as above. This is because normalizing X as above by substracting the column means results in a dense matrix which easily exceeds available memory. Thus the computation is split up suitably to always avoid the creation of dense matrices. Letting v be the column means as above, and $w \in \mathbb{R}^n$ be the column sums,

$$\mathrm{cov}(X) = \frac{1}{m-1}\left(X^\top X - wv^\top - vw^\top + mvv^\top\right).$$

We observe that computing $X^\top X$ involves only the sparse input matrix and one sparse matrix multiplication (which can be done efficiently). The other three terms are vector-vector products resulting in dense $n\times n$ matrices, the (necessary) size of the output covariance matrix.

A.2   Jaccard Similarity Matrix

Denoting the ith column of X as X_i, each entry (i,j) of the Jaccard matrix is given as

$$\mathrm{jac}(X)_{ij} = \frac{|\{k : X_{ik} \wedge X_{jk}\}|}{|\{k : X_{ik} \vee X_{jk}\}|}.$$

For this we assume that X is binary.

Naïvely, we iterate over all the entries of the Jaccard matrix and compute them as given above using binary *and* as well as binary *or* operations. This turned out to be slow in our experiments. The following is a faster way to compute the Jaccard matrix in practice even though the asymptotic runtime is unchanged.

Recall that $w \in \mathbb{R}^n$ denotes the column sums of X. Using sparse matrix multiplication, we compute $Y = X^\top X \in \mathbb{R}^{n\times n}$, which is a dense matrix. Let $Z \in \mathbb{R}^{n\times n}$ be the matrix obtained by adding w to all rows and all columns of $-Y$. Observe that $\mathrm{jac}(X) = Y/Z$, where the matrix division is performed entry-wise. Since we only need one sparse matrix multiplication to compute Y (which can be done efficiently), this approach is computationally very fast. The few other operations on the matrices Y and Z are efficient since both matrices are already of same size as the dense Jaccard output matrix.

Table 1. Theoretical runtimes for computing the four similarity matrices. Runtimes differ between dense and sparse implementations. The parameters are: dimensions $m \in \mathbb{N}$ and $n \in \mathbb{N}$ of the input data $X \in \mathbb{R}^{m\times n}$, matrix sparsity parameter $s \in [0,1]$.

Method	Dense	Sparse
Covariance matrix	$O(mn^2)$	$O(smn^2 + n^2)$
Unweighted Jaccard	$O(mn^2)$	$O(smn^2 + n^2)$
Weighted Jaccard	$O(mn^2)$	$O(smn^2 + mn)$
GRM matrix	$O(mn^2)$	$O(smn^2 + n^2)$

B   Theoretical Runtimes

Table 1 shows the theoretical runtimes for both dense and sparse implementations. As can be seen, the theoretical runtimes for the dense computations are equal, but runtimes for sparse implementations differ. A detailed overview of the computations being carried out and their runtimes is given below. All efforts are given in the dimensions $m \in \mathbb{N}$ and $n \in \mathbb{N}$ of the input data $X \in \mathbb{R}^{m \times n}$, and the matrix sparsity parameter $s \in [0, 1]$ (the proportion of non-zero matrix entries).

Computing the covariance matrix involves calculating the column means of X and substracting them from the matrix X ($O(mn)$ for the dense case and $O(n^2)$ for the sparse case involving a correction with outer products, see Sect. A.1). Multiplying $Y^\top Y$ takes $O(mn^2)$ in dense and $O(smn^2)$ in sparse algebra.

Computing the Jaccard matrix involves calculating $Y = X^\top X$ ($O(mn^2)$ in dense and $O(smn^2)$ in sparse algebra) and adding the column sums of X (computed in $O(mn)$ in dense and $O(smn)$ in sparse algebra) to all rows and columns ($O(n^2)$ in both dense and sparse algebra).

Computing the weighted Jaccard matrix (or s-matrix) involves calculating the row sums of the input matrix which are used as weights ($O(mn)$ in dense and $O(smn)$ in sparse algebra), componentwise multiplication of all columns with the weights ($O(mn)$ in dense and $O(smn)$ in sparse algebra), and one matrix-matrix multiplication ($O(mn^2)$ in dense and $O(smn^2)$ in sparse algebra).

Computing the GRM matrix involves the calculation of population frequencies across rows ($O(mn)$ in dense and $O(smn)$ in sparse algebra), one matrix-matrix multiplication ($O(mn^2)$ in dense and $O(smn^2)$ in sparse algebra), as well as multiplying the input matrix with the population frequencies ($O(mn)$ in dense and $O(smn)$ in sparse algebra). Additionally, one outer vector product is required ($O(n^2)$ in both dense and sparse algebra).

Table 2. Computation of the global eigenvector (global EV) and complete stratification scan of chromosome 1 of the 1,000 Genome Project as a function of the window size. Runtimes in seconds for locStra and PLINK2.

Window size	locStra		PLINK2	
	Global EV	Full scan	Global EV	Full scan
1000	1.4	332.1	65.3	6343.6
10000	1.5	31.3	61.2	731.7
100000	1.5	4.9	66.7	189.1

C   Comparison of locStra to PLINK2

Table 2 shows a runtime comparison between locStra and PLINK2. As test data we use chromosome 1 of the 1,000 Genome Project. Before running either locStra or PLINK2, we prepare the raw data from the 1,000 Genome Project using the same parameters as given in Sect. 3. However, locStra and PLINK2 require different input files, and thus we write out the processed data once in the *.bed* format for PLINK2, and once as *.Rdata* file containing a sparse matrix of class *Matrix* in *R*.

A local stratification scan can be performed in PLINK2: With the command *--pca 1*, the first eigenvector can be computed for an input *.bed* file. In order to do a sliding window scan, we use the parameters *--from* and *--to* followed by the rs numbers to specify a local window. All eigenvectors are written to an output file with extension *.eigenvec* by PLINK2, from which we read the vectors and compute correlations in *R*.

In the locStra package, the local stratification scan is performed using the function *fullscan* as described in Sect. 2.2.

The results in Table 1 show that even for the computation of the single global eigenvector, locStra is considerably faster than PLINK2. All runtimes for both locStra and PLINK2 include the time to read the *.Rdata* or *.bed* input files. For a full scan, PLINK2 needs to (inefficiently) write the eigenvector data for each local window into a file. In comparison to PLINK2, locStra is around one order of magnitude faster, where the speed-up is more pronounced for larger window sizes.

References

1. Bates, D., Eddelbuettel, D.: Fast and elegant numerical linear algebra using the RcppEigen package. J. Stat. Softw. **52**(5), 1–24 (2013)
2. Chang, C.C., Chow, C.C., Tellier, L.C., Vattikuti, S., Purcell, S.M., Lee, J.J.: Second-generation PLINK: rising to the challenge of larger and richer datasets. GigaScience, **4** (2015)
3. Devlin, B., Roeder, K.: Genomic control for association studies. Biometrics **55**(4), 997–1004 (1999)
4. Laird, N.M., Lange, C.: The Fundamentals of Modern Statistical Genetics. SBH. Springer, New York (2011). https://doi.org/10.1007/978-1-4419-7338-2
5. Lee, S., Epstein, M.P., Duncan, R., Lin, X.: Sparse principal component analysis for identifying ancestry-informative markers in genome-wide association studies. Genet. Epidemiol. **36**(4), 293–302 (2012)
6. Martin, E.R., et al.: Properties of global and local ancestry adjustments in genetic association tests in admixed populations. Genet. Epidemiol. **42**(2), 214–229 (2018)
7. Patterson, N., Price, A.L., Reich, D.: Population structure and Eigenanalysis. PLoS Genet. **2**(12), e190 (2006)
8. Price, A.L., et al.: Principal components analysis corrects for stratification in genome-wide association studies. Nat. Genet. **38**, 904–909 (2006)
9. Price, A.L., et al.: Sensitive detection of chromosomal segments of distinct ancestry in admixed populations. PLoS Genet. **5**(6), e1000519 (2009)
10. Pritchard, J.K., Stephens, M., Rosenberg, N.A., Donnelly, P.: Association mapping in structured populations. Am. J. Hum. Genet. **67**(1), 170–181 (2000)
11. Prokopenko, D., et al.: Utilizing the Jaccard index to reveal population stratification in sequencing data: a simulation study and an application to the 1000 Genomes Project. Bioinformatics **32**(9), 1366–1372 (2016)

12. Purcell, S., Chang, C.: PLINK2 (2019)
13. Schlauch, D., Fier, H., Lange, C.: Identification of genetic outliers due to sub-structure and cryptic relationships. Bioinformatics **33**(13), 1972–1979 (2017)
14. Schlauch, D.: Implementation of the stego algorithm - similarity test for estimating genetic outliers (2016)
15. The 1000 Genomes Project Consortium: A global reference for human genetic variation. Nature, **526**, 68–74 (2015)
16. Mises, R.V., PollaczekGeiringer, H.: Praktische Verfahren der Gleichungsaufloesung. ZAMM Zeitschrift fur Angewandte Mathematik und Mechanik **9**, 152–164 (1929)
17. Wang, B., Sverdlov, S., Thompson, E.: Efficient estimation of realized kinship from single nucleotide polymorphism genotypes. Genetics **205**(3), 1063–1078 (2017)
18. Zhong, Y., Perera, M.A., Gamazon, E.R.: On using local ancestry to characterize the genetic architecture of human traits: genetic regulation of gene expression in multiethnic or admixed populations. Am. J. Hum. Genet. **104**(6), 1097–1115 (2019)

A New Graph Database System for Multi-omics Data Integration and Mining Complex Biological Information

Ishwor Thapa(iD) and Hesham Ali[⊠](iD)

College of Information Science and Technology,
University of Nebraska at Omaha, Omaha, USA
{ithapa,hali}@unomaha.edu

Abstract. Due to the advancement in high throughput technologies and robust experimental designs, many recent studies attempt to incorporate heterogeneous data obtained from multiple technologies to improve our understanding of the molecular dynamics associated with biological processes. Currently available technologies produce wide variety of large amount of data spanning from genomics, transcriptomics, proteomics, and epigenetics. Due to the fact that such multi-omics data are very diverse and come from different biological levels, it has been a major research challenge to develop a model to properly integrate all available and relevant data to advance biomedical research. It has been argued by many researchers that the integration of multi-omics data to extract relevant biological information is currently one of the major biomedical informatics challenges. This paper proposes a new graph database model to efficiently store and mine multi-omics data. We show a working model of this graph database with transcriptomics, genomics, epigenetics and clinical data for three cancer types from the Cancer Genome Atlas. Moreover, we highlight the usefulness of graph database mining to extract relevant biological interpretations and also to find novel relationships between different data levels.

Keywords: Graph database · Data integration · Multi-omics data · Information mining

1 Introduction

Molecular function of a cell can be traced back to its cellular components and processes. These processes can be regulated at multiple levels through different cellular mechanisms such as mutations and copy number aberration in the genomic landscape, DNA methylation in epigenetics and gene expression in transcription. Due to the advancement in genomic and transcriptomic technology platforms such as SNP Array, RNASeq and Methylation Assay, a single study can have multitude of data types. Large-scale studies with heterogeneous data types have been successfully implemented. The Cancer Genomic Atlas (TCGA)

© Springer Nature Switzerland AG 2020
I. Măndoiu et al. (Eds.): ICCABS 2019, LNBI 12029, pp. 171–183, 2020.
https://doi.org/10.1007/978-3-030-46165-2_14

is a comprehensive resource for cancer genomic profiles [24]. Among many data types, TCGA has data for clinical, image, DNA sequencing, miRNA sequencing, mRNA expression, DNA methylation and Copy Number information for more than 30 cancer types. Similar multi-level datasets can be found in the Cancer Cell Line Encyclopedia (CCLE), which compiles gene expression, chromosomal copy number and sequencing data together with pharmacological profiles of 24 anticancer drugs [4]. At a smaller scale, many studies have been carried out to integrate genomics, transcriptomics, proteomics, and metabolomics in an individual experimental setting [11,17,21]. As the biomedical research community continues to produce more multi-omics data, we will need sophisticated methodologies to mine useful information from the integrated data source.

1.1 Integration of Multi-omics Data

Due to high dimensionality in each data level, dependencies across the layers in multi-level data, and the differences in technologies, it is challenging to integrate multi-omics data effectively. Many general-purpose data exploration methods such as finding correlations across data levels, extracting relevant information at different levels using principal component analysis or independent component analysis have been used for multi-omics data integration. The MethyCancer database provides MethyView and Methy&Cancer, visualization tools to study the interplay between DNA methylation, gene expression and cancer [10]. The cBioPortal is a web portal to explore, visualize and analyze multidimensional cancer genomics data [9]. The GenomeSnip is a web-based visual analytics platform for exploration of human genome and its relationship to other features [6]. The canEvolve web portal provides visualization and analysis tool for integrative oncogenomics [22]. The GenomeSnip platform provides an interactive visualization tool that allows genome exploration and retrieval of different relationships of genomic features [6]. For systematic integration of the heterogeneous data types, multivariate analysis such as (multiple) co-inertia analysis has been applied in a number of studies [8] and [16].

1.2 Graph Database

Recently, newer studies have been proposed to use a graph database for integration of heterogeneous biological data [3,23,27]. A graph database is a database management system, which allows create, read, update and delete (CRUD) methods to be performed on a graph data model. Unlike other database systems where the relationships are inferred from properties such as foreign keys, the graph data model defines the relationships as equally important to nodes/entities [20]. Neo4J is an open source graph database management system that is ACID (Atomicity, Consistency, Isolation, Durability) compliant [20]. Yoon et al. have compared MySQL and Neo4J and showed an increase in performance level while using a graph database for complex biological relationships [27]. A number of recent studies have used Neo4J as their graph database in different biological contexts. Balaur et al. have implemented Neo4J database in order to model human metabolic network [2] and epigenetic and genetic interdependencies in colon cancer [3]. Fabregat et al. presented Neo4J graph database to facilitate easy traversal and knowledge discovery and thereby reducing the average query time by 93% [7]. More studies such as [5, 13, 25] show feasibility of using graph database and Neo4J while modeling different biological data sources.

In this paper, we show how multi-omics data can be stored as a graph model in a graph database such as Neo4J. We also highlight the queries applied to this database and their corresponding results that show more relevant biological information than given by the existing databases for cancer genomics. We also discuss the efficacy of this database in discovering novel relationships. The graph database contains transcriptomics, genomics, epigenetics and clinical data from multiple cancer types. We implemented the graph database in Neo4J to store, analyze and query multi-omics data from the cancer samples and used pattern-matching queries written in Cypher language.

2   Method

This section describes how the graph database for multi-omics data is created with an example dataset from the Genomic Data Commons (GDC), formerly called the TCGA data atlas. Although we utilize data from only three cancer types in this paper, the methodology described in this section can be applied to create a comprehensive graph database for all types of cancer for which the data is available in GDC. The version of the GDC data release is 10.1.

2.1   Dataset

For this study, we selected transcriptomics (gene expression), epigenetics (methylation) and genomics (mutation) data from GDC, for three forms of cancer, viz. breast cancer (BRCA), prostate adenocarcinoma (PRAD) and the pancreatic adenocarcinoma (PAAD). Additionally, we obtained the clinical information for all of the samples in these three groups. Table 1 shows the summary about the dataset used for this study. Using the GDC Data Portal, we apply GDC queries such as shown in Listing 1.1 to obtain the *manifest* files and then download the files for all four types of data (gene expression, methylation, mutation and clinical) using the *gdc-client* tool. The *manifest* file is a table containing metadata such as unique identifier, filename, md5 hash, size and the status of the files to be downloaded.

```
files.analysis.workflow_type in ["HTSeq - Counts"] and files.data_category in ["
  ↪ Transcriptome Profiling"] and files.access = "open" and cases.project.project_id
  ↪ = "TCGA-PRAD"
```

Listing 1.1. GDC query for transcriptomics data for prostate cancer samples

2.2   Data Pre-processing

In order to store meaningful data from the 7409 files from 12 of the data sources (see Table 1), we pre-process all the files to extract and/or reformat the data. For the methylation data type, only the genes that have been hyper or hypo-methylated are extracted. We used the threshold $\beta > 0.9$ for hyper-methylation and $\beta < 0.1$ for hypo-methylation found in the *island* region. For the transcriptomics data type, all the *htseq* count values are normalized to $log2CPM$ (Counts Per Million) values and the *ENSG IDs* are converted to gene symbols using the annotation file provided by the GDC. For the mutation data type, only the missense mutations (resulting into different amino acid) are extracted. From the clinical files, all of the data is stored in the database.

Table 1. Summary of the dataset used for this study

Data type	No. of samples:BRCA	No. of samples:PRAD	No. of samples:PAAD
Clinical	1097	500	184
Methylation	1234	553	194
Gene expression	1234	553	194
Mutation	986	498	182

2.3   The Proposed Graph Database for *omics* Data

We use the *Property Graph Model*, a variant of graph modeling, to capture the entities and the relationships in our data. The property graph model requires the relationships to be named and directed with a start and end node. Figure 1 shows entities and the relationships in our graph database, both of them containing key-value pair properties. Nodes (entities) are represented in circles and the relationships are represented by the edges. Individual case (patient) in the TCGA data can have multiple samples. Sometimes, there is expression data from both normal and tumor tissues for the same patient. If there was a follow-up visit and the samples were collected again, there can be multiple samples for a single case. Hence, we have *Case* and *Sample* as two different nodes connected by *HAS_SAMPLE* relationship (see Fig. 1). This relationship has a property called *sample_type* that describes whether the sample is from tumor or normal tissues. All of the clinical information about the patient is stored as the node property of the *Case* as shown in Table 2. Each cancer type shares some clinical properties to other cancer types but it may also have unique subset of clinical information about the patient. Hence, the property keys are not always the same for all of the *Case* nodes. Unlike relational databases where the attributes of an entity are fixed, the property graph model allows to have different property keys for the same entity class. This is very much useful especially in storing data from different diseases because clinical information for patients with different diseases have very unique measures. For instance, *'her2 receptor status'* is specific to breast cancer dataset to represent if sample has positive or negative role of 'her2'. Next, all of the transcriptomics, methylation and mutation information is summarized at the gene level; therefore, we have *Gene* as a node. For each *Sample-Gene* pair, the gene expression value that is captured by the normalized *htseq* count (log2 count of reads per million mapped reads) is stored as a relationship property between the *Sample-Gene* pair (see Fig. 1). For genes with multiple transcripts, multiple relationships are created with different ENSG ids and expression values as their relationship properties. The hyper/hypo-methylation and mutation events are stored as distinct relationships between a *Sample-Gene* pair with the chromosome number and the location as the relationship properties (see Table 3).

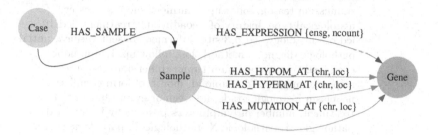

Fig. 1. Graph database for transcriptomics, methylation and mutation data

To implement above described property graph model, we used Neo4J community edition of version 3.2.0 as our database. The entities (nodes), relationships and their properties are imported into the Neo4J database using *neo4j-import* tool. We have also written Python programs using *Neo4J Python driver* to assign and update property values of the nodes.

2.4   Data Mining from Graph Database

Graph databases also provide graph based query languages to extract information from the database. Neo4J provides query language called *Cypher* for this purpose. A Cypher query contains a pattern that is searched in the graph and returns a sub-graph with the input pattern and local matches. Example cypher queries are shown in Listing 1.2 where the first query can be used to query for nodes (*Case*) with specific property values such as disease code and pathologic stage. In the second example, query for a path from *Case* to *Sample* that contains the *HAS_SAMPLE* relationship is being shown. In these examples, the search pattern has specific nodes and relationships to look for. However, cypher queries can deal with anonymous nodes or relationships by simply removing their labels from the query. More complex queries will be illustrated in the results section. Neo4J also provides *Neo4J Browser* for querying the database and visualization of the results. The cypher queries such as mentioned in Listing 1.2 can be run

Table 2. Entities (Nodes) and their properties in the graph database (the *Case* entity property differ slightly between different cancer types)

Node/Entity	Node properties
Case	adenocarcinoma invasion, age at initial pathologic diagnosis, anatomic neoplasm subdivision, batch number, bcr, bcr drug barcode, bcr drug uuid, bcr followup barcode, bcr followup uuid, bcr p, atient barcode, bcr patient uuid, caseId, day of dcc upload, day of form completion, days to birth, days to death, days to drug therapy end , days to drug therapy start , days to initial pathologic diagnosis, disease code, drug name, ethnicity, family history of cancer, file uuid, followup case report form submission reason, followup treatment success, gender, histological type, history of neoadjuvant treatment, icd_10, icd_o_3 histology, icd_o_3 site, informed consent verified, initial pathologic diagnosis method, lost follow up, lymph node examined count, maximum tumor dimension, measure of response, month of dcc upload, month of form completion, neoplasm histologic grade, new tumor event after initial treatment, number of lymphnodes positive by he, other dx, pathologic M, pathologic N, pathologic T, pathologic stage, patient barcode, patient id, person neoplasm cancer status, primary lymph node presentation assessment, primary therapy outcome success, project code, race, radiation therapy, residual tumor, source of patient death reason, surgery performed type, system version, targeted molecular therapy, therapy ongoing, therapy type, tissue prospective collection indicator, tissue retrospective collection indicator, tissue source site , tobacco smoking history, tumor tissue site, tumor type, tx on clinical trial, vital status, year of dcc upload, year of form completion, year of initial pathologic diagnosis
Gene	"Symbol", "geneId"
Sample	"sample_barcode", "sampleId"

Table 3. Relationships and their properties in the graph database

Relationships	Relationship properties
HAS_EXPRESSION	ENSG, normalized_count
HAS_HYPERM_AT	Chromosome, location
HAS_HYPOM_AT	Chromosome, location
HAS_MUTATION_AT	chromosome, location
HAS_SAMPLE	sample_type

in the Neo4J Browser to visualize the result and to export it into other formats
such as *CSV or PNG.*

```
MATCH (c:Case) WHERE c.disease_code='PAAD' AND c.pathologic_stage="Stage IIA" RETURN c
MATCH r= (:Case)-[:HAS_SAMPLE]->(:Sample) RETURN r
```

Listing 1.2. Sample Cypher queries

3 Results

We next highlight the graph database created using the clinical information, gene
expression, methylation and mutation data and showcase several graph-based
queries and their results from the graph database. Furthermore, we illustrate
how novel relationships can be mined from the integrated database.

3.1 The Integrated Graph Database

The graph database containing gene expression, methylation, mutation and clinical data for samples from the three types of cancer (BRCA, PRAD and PAAD)
consisted of 62,232 nodes and 268,018,177 relationships. The Table 4 shows exact
number of nodes and relationships for each of their types. The database takes
19 GB of disk space.

Table 4. Number of nodes and relationships in the database

Node	counts	Relationship	counts
Case	1781	HAS_EXPRESSION	113,854,650
Sample	2016	HAS_HYPERMETH_AT	15,606,784
Gene	58,435	HAS_HYPOM_AT	138,458,403
		HAS_MUTATION_AT	98,340

3.2 Querying the Database

Since the graph database can be queried for nodes and relationships, we show
results of few cypher queries, which only involve the nodes as well as those with
nodes and relationships.

Demographic Queries. Following two queries listed in Listing 1.3 show how
nodes with specific property values can be obtained from the graph database.
We show two important demographics from the PAAD dataset, viz. (a) number
of patients for each stage and (b) number of patients per race. The Table 5 shows
the results from these queries.

```
match (c:Case) where c.disease_code="PAAD" return c.pathologic_stage,count(c.caseId) order
    ↪ by c.pathologic_stage
match (c:Case) where c.disease_code="PAAD" return c.race,count(c.patient_barcode)
```

Listing 1.3. Cypher query for number of patients for different stages in PAAD

Table 5. Result of queries related to demographics (left: pathologic stage, right: race)

c.pathologic_stage	count(c.caseId)
Stage I	1
Stage IA	5
Stage IB	15
Stage IIA	30
Stage IIB	122
Stage III	4
Stage IV	5
null	3

c.race	count(c.patient_barcode)
Asian	11
African American	7
White	162
null	5

Gene Expression of Selected Genes in Prostate Cancer Across Multiple Sample Types. Next we show how the graph database can be used to extract information on how the gene expression values for a subset of genes vary across different sample types such as normal tissue, primary tumor and metastatic. We choose the PRAD dataset in this case and show the genes *TSPAN13* and *MCM7* both have higher expression in tumor samples than in normal (see Table 6). The authors of [19] and [1] have individually shown the overexpression of *MCM7* and *TSPAN13* genes in prostate cancer.

```
MATCH (c:Case{disease_code:"PRAD"}) -[hs:HAS_SAMPLE]->(s:Sample) -[e:HAS_EXPRESSION]->(g:
  ↪ Gene) where g.symbol in ['TSPAN13','MCM7'] return hs.sample_type, avg(toFloat(e.
  ↪ ncount)),g.symbol
```

Listing 1.4. Cypher query for gene expression values in different tissue types in PRAD

Frequently Mutated Genes in Each Cancer Type. Next we show how top ten mutation events and the genes with those frequent mutations can be obtained from the graph database. The Listing 1.5 shows the Cypher query used to generate the frequently mutated list of genes and the mutation counts

Table 6. Result from query in Listing 1.4 ($n_{PrimaryTumor} = 517$, $n_{SolidTissueNormal} = 52$, $n_{Metastatic} = 1$)

hs.sample_type	avg (toFloat(e.ncount))	g.symbol
Metastatic	8.22	TSPAN13
Primary Tumor	7.512208835	TSPAN13
Solid Tissue Normal	6.5775	TSPAN13
Metastatic	7.23	MCM7
Primary Tumor	6.207148594	MCM7
Solid Tissue Normal	5.795576923	MCM7

presented in Table 7. Most of these genes are already known to have frequent mutations. In Sect. 5, we discuss further regarding these genes from this result and highlight its significance. The *TTN* gene being the longest gene in human genome is more frequently mutated in all three cancer types.

Table 7. Result from query in Listing 1.5

Cancer	Gene	Frequency	Cancer	Gene	Frequency	Cancer	Gene	Frequency
	PIK3CA	340		KRAS	140		TTN	62
	TTN	258		TTN	75		SPOP	54
	TP53	204		TP53	72		TP53	39
	MUC16	137		MUC16	33		MUC16	32
BRCA	RYR2	73	PAAD	USH2A	22	PRAD	SYNE1	28
	USH2A	62		RYR3	20		SPTA1	24
	HMCN1	61		SYNE1	20		OBSCN	22
	FLG	60		LRP1B	17		LRP1B	20
	SYNE1	59		COL5A1	17		RYR2	16
	DMD	53		RYR1	17		RYR1	16

```
match p=(c:Case {disease_code:"BRCA"}) -[hs:HAS_SAMPLE]->(s:Sample)-[e:HAS_MUTATION_AT]->(g
  ↪ :Gene) return g.symbol,count(e.ncount) as ec order by ec desc
```
Listing 1.5. Cypher query for finding top 10 mutation events in patients with BRCA

Sub-network of Samples and Genes Mutated in Stage IV Patients in PAAD. We queried the graph database to find genes that are mutated in at least two samples in stage IV patients in PAAD. The Fig. 2 shows a sub-graph containing the genes (green nodes), *KRAS* and *TP53*, which have mutations in at least two patients (pink nodes) in Stage IV of PAAD. Furthermore, the *KRAS* mutation is observed in chromosome 12 at position 25245350 for all five samples while the *TP53* mutation in three of the samples are in three different locations.

Methylation Status of Specific Genes from Pancreatic Cancer Samples. Next, we show how the integrated graph database can be used to extract epigenetic events relevant to cancer. Specifically, we use pancreatic cancer samples and find out number of patients who have hyper-methylation events in previously known hyper-methylated genes. We used *CACNA1G*, *TIMP2* and *RUNX3* genes, previously shown to have hyper-methylation activity in pancreatic cancer, to identify PAAD samples from our dataset with hyper-methylation events in these genes [18,26]. We show that a large number of samples (out of 184 total) have hyper-methylation in these three genes (see Table 8).

```
match p=(c:Case {disease_code:"PAAD"}) -[hs:HAS_SAMPLE{sample_type:"Primary Tumor"}]->(s:
  ↪ Sample)-[e:HAS_HYPERM_AT]->(g:Gene) where g.symbol in ['CACNA1G','TIMP2','RUNX3']
  ↪ return g.symbol, count(distinct(c.patient_barcode))
```
Listing 1.6. Cypher query for finding count of patients with hyper methylation

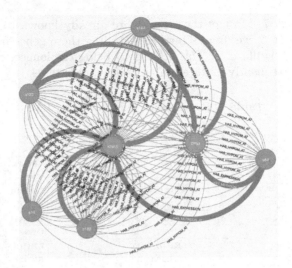

Fig. 2. Resulting subgraph finding genes mutated in stage IV patients in PAAD

Table 8. Result from query in Listing 1.6 to count the number of samples with hyper-methylation events in PAAD dataset ($n = 184$)

Gene	Count of patients with Hyper Methylation in CpG islands
CACNA1G	139
RUNX3	184
TIMP2	182

Mutation Count and Different Pathologic Status. To examine if the mutation count is increased in advanced stages of cancer compared to early stages, we used the graph database to query for number of patients and the number of mutations in each stage of cancer from breast cancer dataset. We aggregated the results from different stages within each stage. For instance, we combined the numbers from Stage I, Stage IA, Stage IB and represented as Stage I (see Table 9). It was observed that the average mutation per patient in Stage

Table 9. Average number of mutations per stage in BRCA dataset

Stages	Count of mut	Count of cases	Average mut
Stage I (combined)	4664	93	50.15
Stage II (combined)	39813	617	64.53
Stage III (combined)	9746	248	39.30
Stage IV	1667	20	83.35

IV is the highest among all four stages. However, the trend across all stages doesn't show a clear pattern.

4   Availability

The source code, a demo dataset and basic instructions on how to populate the graph database and run database server are available in GitHub repository at the URL, https://github.com/ishworthapa/graphd. The source code requires Python3 and the instructions are written for Debian based Linux distributions. The Neo4J database server that is being utilized is the community edition.

5   Discussions

We presented a graph database that integrates the mutation, DNA methylation, gene expression and clinical information from cancer patients in the cancer genome atlas (TCGA). Previous studies such as [27] and [7] have shown an increase in performance while using a graph database for biological data as compared to traditional relational database. The Neo4J graph database and its query language, Cypher provided a strong platform for efficient storage and access to heterogeneous data from multi-omics studies. Comparing the performance of property graph model with other graph models such as resource description framework (RDF) can be a future step. We highlighted several powerful and unique example queries and results that are relevant for cancer researchers and not as easily available in existing platforms. Most of the results we showed were discovered in previous literatures. However, we want to highlight that by querying to this database and analyzing the query results, researchers can formulate novel data-driven hypothesis. The mutations in *LRP1B* gene that have been shown to be associated with hepatocellular, lung and other cancer forms was found to be frequently mutated in PRAD and PAAD in our study (see Table 7) [12,15]. Similarly we found that the *RUNX3* gene expression is decreased from normal tissue (5.1325) to primary tumor (3.65) to metastatic tissue (2.06) using a query similar to Listing 1.4, where we changed the gene name to *RUNX3*. We also note that the *RUNX3* gene is frequently hyper-methylated in PAAD (see Table 8). It has been shown that hyper-methylation of RUNX3 down-regulates its gene expression in other cancers. The relationship between multiple data levels such as hyper-methylation and down-regulation of gene expression can be uniquely studied using our graph database. Moreover, similar studies can explore the role of hyper-methylation in *TIMP2*, another frequently hyper-methylated gene in PAAD cancer.

In future, more data types such as copy number aberration and miRNA sequencing data, which are available with open access in TCGA, will also be incorporated into the graph database. The *Gene* node can be expanded with gene lengths that can help in generating tumor mutational burden. To make this graph database richer, information from other databases such as *MutSig* can be added for each mutation site [14]. In addition, we intend to include data from all forms of cancer that are available in TCGA and make the platform publicly available. We envision that this platform will allow cancer researchers to widely use publicly available data from TCGA with a holistic view from multiple heterogeneous data sources.

Acknowledegment. This work was partly funded by the System Science Grant supported by Nebraska Research Initiative (NRI).

References

1. Arencibia, J.M., Martín, S., Pérez-Rodríguez, F.J., Bonnin, A.: Gene expression profiling reveals overexpression of TSPAN13 in prostate cancer. Int. J. Oncol. **34**(2), 457–463 (2009)
2. Balaur, I., Mazein, A., Saqi, M., Lysenko, A., Rawlings, C.J., Auffray, C.: Recon2Neo4j: applying graph database technologies for managing comprehensive genome-scale networks. Bioinformatics **33**(7), 1096–1098 (2016)
3. Balaur, I., et al.: Epigenet: a graph database of interdependencies between genetic and epigenetic events in colorectal cancer. J. Comput. Biol. **24**(10), 969–980 (2017)
4. Barretina, J., et al.: The cancer cell line encyclopedia enables predictive modelling of anticancer drug sensitivity. Nature **483**(7391), 603 (2012)
5. Costa, R.L., Gadelha, L., Ribeiro-Alves, M., Porto, F.: Gennet: An integrated platform for unifying scientific workflow management and graph databases for transcriptome data analysis, p. 095257. bioRxiv (2016)
6. Decker, S., Deus, H., Iqbal, A., Kamdar, M., Saleem, M.: Genomesnip: fragmenting the genomic wheel to augment discovery in cancer research. In: Conference on Semantics in Healthcare and Life Sciences (CSHALS). ISCB (2014)
7. Fabregat, A., et al.: Reactome graph database: efficient access to complex pathway data. PLoS Comput. Biol. **14**(1), e1005968 (2018)
8. Fagan, A., Culhane, A.C., Higgins, D.G.: A multivariate analysis approach to the integration of proteomic and gene expression data. Proteomics **7**(13), 2162–2171 (2007)
9. Gao, J., et al.: Integrative analysis of complex cancer genomics and clinical profiles using the cbioportal. Sci. Signal. **6**(269), pl1–pl1 (2013). https://doi.org/10.1126/scisignal.2004088. http://stke.sciencemag.org/content/6/269/pl1
10. He, X., et al.: Methycancer: the database of human dna methylation and cancer. Nucleic Acids Res. **36**(suppl-1), D836–D841 (2007)
11. Hirai, M.Y., et al.: Integration of transcriptomics and metabolomics for understanding of global responses to nutritional stresses in arabidopsis thaliana. Proc. Nat. Acad. Sci. USA **101**(27), 10205–10210 (2004)
12. Kan, Z., et al.: Whole-genome sequencing identifies recurrent mutations in hepatocellular carcinoma. Genome Res. **23**(9), 1422–1433 (2013)
13. Kazantsev, F., et al.: Mammoth: a new database for curated mathematical models of biomolecular systems. J. Bioinform. Comput. Biol. **16**(01), 1740010 (2018)
14. Lawrence, M.S., et al.: Mutational heterogeneity in cancer and the search for new cancer-associated genes. Nature **499**(7457), 214 (2013)
15. Lee, W., et al.: The mutation spectrum revealed by paired genome sequences from a lung cancer patient. Nature **465**(7297), 473 (2010)
16. Meng, C., Kuster, B., Culhane, A.C., Gholami, A.M.: A multivariate approach tothe integration of multi-omics datasets. BMC Bioinform. **15**(1), 162 (2014). https://doi.org/10.1186/1471-2105-15-162

17. Network, C.G.A.R., et al.: Comprehensive molecular profiling of lung adenocarcinoma. Nature **511**(7511), 543 (2014)
18. Nomoto, S., et al.: Adverse prognosis of epigenetic inactivation in runx3 gene at 1p36 in human pancreatic cancer. Br. J. Cancer **98**(10), 1690 (2008)
19. Ren, B., et al.: MCM7 amplification and overexpression are associated with prostate cancer progression. Oncogene **25**(7), 1090 (2006)
20. Robinson, I., Webber, J., Eifrem, E.: Graph Databases. O'Reilly Media, Inc., Newton (2013)
21. Romero, R., et al.: The use of high-dimensional biology (genomics, transcriptomics, proteomics, and metabolomics) to understand the preterm parturition syndrome. BJOG Int. J. Obstet. Gynaecol. **113**(s3), 118–135 (2006)
22. Samur, M.K., et al.: canEvolve: a web portal for integrative oncogenomics. PLOS ONE **8**(2), 1–10 (2013). https://doi.org/10.1371/journal.pone.0056228
23. Swainston, N., et al.: biochem4j: Integrated and extensible biochemical knowledge through graph databases. PLoS One **12**(7), e0179130 (2017)
24. Tomczak, K., Czerwińska, P., Wiznerowicz, M.: The cancer genome atlas (TCGA): an immeasurable source of knowledge. Contemp. Oncol. **19**(1A), A68 (2015)
25. Touré, V., et al.: STON: exploring biological pathways using the SBGN standard and graph databases. BMC Bioinform. **17**(1), 494 (2016). https://doi.org/10.1186/s12859-016-1394-x
26. Ueki, T., et al.: Hypermethylation of multiple genes in pancreatic adenocarcinoma. Cancer Res. **60**(7), 1835–1839 (2000)
27. Yoon, B.H., Kim, S.K., Kim, S.Y.: Use of graph database for the integration of heterogeneous biological data. Genomics Inform. **15**(1), 19–27 (2017)

SMART2: Multi-library Statistical Mitogenome Assembly with Repeats

Fahad Alqahtani[1,2] and Ion Măndoiu[1(✉)]

[1] Computer Science and Engineering Department, University of Connecticut,
Storrs, Mansfield, CT, USA
{fahad.alqahtani,ion.mandoiu}@uconn.edu
[2] National Center for Artificial Intelligence and Big Data Technology,
King Abdulaziz City for Science and Technology, Riyadh, Saudi Arabia

Abstract. SMART2 is an enhanced version of the SMART pipeline for mitogenome assembly from low-coverage whole-genome sequencing (WGS) data. Novel features include automatic selection of the optimal number of read pairs used for assembly and the ability to assemble multiple sequencing libraries when available. SMART2 succeeded in generating mitochondrial sequences for 26 metazoan species with WGS data but no previously published mitogenomes in NCBI databases. The SMART2 pipeline is publicly available via a user-friendly Galaxy interface at https://neo.engr.uconn.edu/?tool_id=SMART2.

Keywords: Mitogenome assembly · Multi-library assembly · Low-coverage sequencing

1 Introduction

Mitochondria are cellular organelles present with very rare exceptions in all eukaryotic cells. In most animals, the mitochondria have their own genome, a double-stranded circular DNA molecule typically ranging in size between 15–20 Kb that encodes 37 genes (2 ribosomal RNA genes, 13 protein coding genes, and 22 transfer RNA genes). The mitochondrial genome is inherited maternally, and has much higher copy number than the nuclear genome [24]. The small size, high copy number, and the presence of both coding and regulatory regions that mutate at different rates make the mitochondrial genome an ideal genetic marker. Indeed, mitochondrial sequences have been used in applications ranging from maternal ancestry inference and tracing human migrations [6] to forensic analysis [19]. The mitochondrial DNA has also become the workhorse of biodiversity studies since many non-model species do not have yet a sequenced nuclear genome [12,16].

To date, most such biodiversity studies have been based on sequencing a single gene fragment, such as the Cytochrome C oxidase I (COI) gene, which has been adopted as the preferred "barcode of life" [14,21]. Recently there have been a renewed appreciation for the improved accuracy of taxonomic and phylogenetic analyses performed based on complete mitogenome sequences assembled from

© Springer Nature Switzerland AG 2020
I. Măndoiu et al. (Eds.): ICCABS 2019, LNBI 12029, pp. 184–198, 2020.
https://doi.org/10.1007/978-3-030-46165-2_15

low coverage whole genome shotgun (WGS) reads generated using next generation sequencing (NGS) technologies. Indeed, full length mitogenome sequences capture evolutionary events such as genome rearrangements that are missed in single gene analyses [18]. Furthermore, the exponential decrease in NGS costs has led to an explosion in the number of WGS datasets generated from non-model organisms. For mammals alone, there are currently over two hundred species with paired-end WGS data available in the NCBI SRA database but for which no complete mitogenome is available. Recent studies have also demonstrated that WGS data of sufficient depth for reconstructing mitogenomes can be generated from preserved museum specimens [23], making the approach applicable to rare or even extinct species.

Leveraging the available WGS datasets to expand the number of complete mitogenomes requires bioinformatics pipelines that can assemble and annotate high-quality mitogenomes quickly and with minimal human intervention. Unfortunately, standard genome assemblers often fail to generate high quality mitochondrial genome sequences due to the large difference in copy number between the mitochondrial and nuclear genomes [13]. This has led to the development of specialized tools for reconstructing mitochondrial genomes from WGS data, mainly falling within three categories. *Reference-based* methods such as MTool-Box [8] require the mtDNA sequence of the species of interest or a closely related species, which are often not available for the less-studied species of interest in biodiversity studies. *Seed-and-extend* tools such as MITObim [13] and NOVOPlasty [11] use a greedy approach to extend available seed sequences such as the COI but can have difficulty handling repetitive regions present in some mitochondrial genomes [16]. Finally, *de novo* methods such as Norgal [2] and plasmidSPAdes [7] use coverage-based filtering to remove nuclear WGS reads before performing assembly using the de Bruijn graph of remaining reads.

In [5] we introduced a hybrid method called *Statistical Mitogenome Assembly with RepeaTs* (SMART), which uses a seed sequence to estimate the mean and standard deviation of mtDNA k-mer counts, then positively selects reads with k-mer counts falling within three standard deviations of the estimated mean before performing *de novo* assembly. Experiments in [3] show that for low-depth WGS datasets the positive selection approach implemented by SMART yields higher enrichment for mtDNA reads than the negative selection of Norgal. Furthermore, SMART was shown to produce complete circular mitogenomes with a higher success rate than both seed-and-extend tools MITObim and NOVOPlasty and *de novo* assemblers Norgal and plasmidSPAdes.

In this paper we present an extension of the SMART pipeline, referred to as SMART2, that can take advantage of multiple sequencing libraries when available and automatically selects the optimal number of read pairs used for assembly. We also present experimental results comparing read filtering and assembly accuracy of SMART2 with that of existing state-of-the-art tools, along with the results of a pilot "orphan mitogenomes" project in which SMART2 was used to generate 15 complete and 11 partial mitogenomes for 26 mammals and amphibians without previously published mitogenomes. All novel mitogenomes have been submitted to GenBank as Third Party Annotation (TPA) sequences [9].

2   Methods

The SMART2 pipeline is deployed using a customized instance of the Galaxy framework [1] and is publicly available via a user-friendly Galaxy interface at https://neo.engr.uconn.edu/?tool_id=SMART2 (see Fig. 1). The pipeline was designed for processing paired-end reads in fastq format from one or two WGS libraries. In addition to fastq files, the user specifies the sample name and a seed sequence in fasta format. By default the number of reads is selected automatically as described below, but the user can override the default and manually specify it. Advanced options also allow the user to change the default choices for the number of bootstrap samples (default is 1), k-mer size (default is 31), number of threads (default is 16), and the genetic code used for MITOS annotation (default is the vertebrate mitochondrial code).

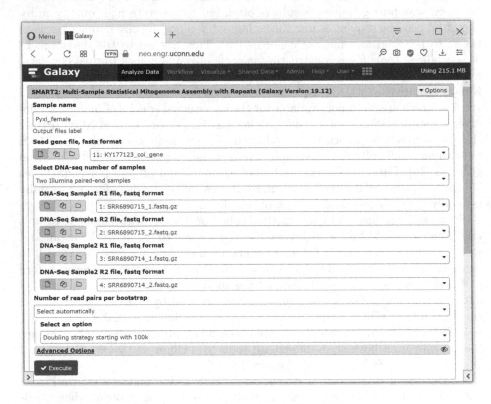

Fig. 1. Galaxy interface of SMART2.

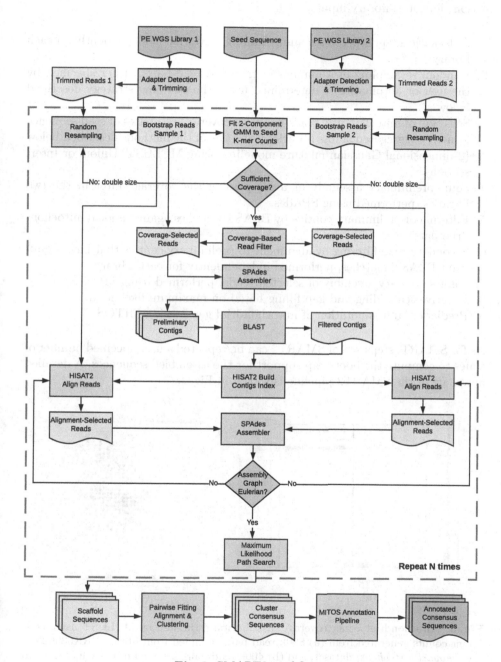

Fig. 2. SMART2 workflow.

The main steps of the SMART2 pipeline follow those of SMART with adaptations for multi-library inputs:

1. Automatic adapter detection and trimming, performed independently for each library.
2. Random resampling of a number of trimmed read pairs, either specified by the user or automatically determined using the doubling strategy described below.
3. Selection of mitochondrial reads based on coverage estimates of seed sequence k-mers – aggregated across libraries using one of the methods described below (2-dimensional Gaussian mixture modeling using MCLUST, Union, or Intersection).
4. Joint preliminary assembly of reads passing the coverage filter in the two libraries, performed using SPAdes.
5. Filtering of preliminary contigs by BLAST searches against a local mitochondrial database.
6. Secondary read filtering by alignment to preliminary contigs that have significant BLAST matches, performed independently for each library.
7. Joint secondary assembly of selected reads, performed using SPAdes.
8. Iterative scaffolding and gap filling based on maximum likelihood.
9. Prediction and annotation of mitochondrial genes using MITOS.

As for SMART, steps 2–8 of SMART2 can be repeated a user-specified number of times to compute the bootstrap support for the assembled sequences. A detailed flowchart of the SMART2 pipeline is shown in Fig. 2.

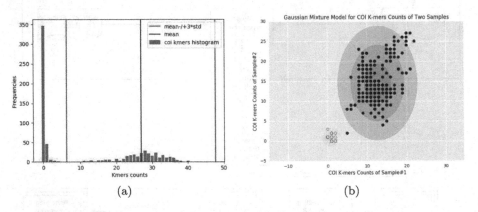

(a)   (b)

Fig. 3. Mitochondrial k-mer coverage distribution estimated by MCLUST using seed k-mer counts generated from (a) 800k read pairs sampled from library SRR630623 of the *Anopheles stephensi* dataset, and (b) 400k read pairs sampled from each of the two libraries of the *Anopheles stephensi* dataset.

2.1 Coverage-Based k-mer Classification

For a single library SMART2 uses the same method as SMART for classifying k-mers as mitochondrial or nuclear in origin. Specifically, SMART2 uses MCLUST [22] to fit a two-component Gaussian mixture model to the one-dimensional distribution of counts of seed sequence k-mers. The upper component of the fitted model is taken as a proxy for the corresponding mtDNA k-mer count distribution, and all k-mers that have a count within 3 standard deviations of the estimated upper component mean are classified as mitochondrial (see Fig. 3(a)).

For two libraries the natural extension of this approach would be to fit a two-component Gaussian mixture model to the *two-dimensional* distribution of counts of seed sequence k-mers (see Fig. 3(b)). Unfortunately experimental results in Sect. 3 show that this approach (referred to as "MCLUST") has relatively poor read filtering performance. Consequently, we implemented in SMART2 two alternative approaches for k-mer classification. Both rely on first independently classifying each k-mer as mitochondrial or nuclear based on fitting two-component Gaussian mixture models to the one-dimensional distributions of counts of seed sequence k-mers of each library. The "Union" method ultimately classifies a k-mer as mitochondrial if it is classified as such based on either one of the libraries, while the "Intersection" method does so if the k-mer is classified as mitochondrial according to *both* libraries.

Table 1. Multi-library WGS datasets with published mtDNA sequences.

Species	Library ID	Read length	% mtDNA	Seed ID	Seed length	Reference ID	Reference length
A. stephensi	SRR630623	2 × 101	0.041%	MK726121	704	KT899888	15,371
	SRR630669	2 × 101	0.041%				
A. funestus	SRR630620	2 × 101	0.03%	MK300232	709	MG742199	15,349
	SRR630619	2 × 101	0.032%				
D. mauritiana	SRR1560275	2 × 76	1.033%	HM630860	560	AF200830	14,964
	SRR1560276	2 × 76	1.120%				
P. major	SRR2961765	2 × 100	0.313%	GQ482300	694	NC_040875	16,777
	SRR2961767	2 × 100	0.313%				
P. humilis	SRR765709	2 × 101	0.215%	EU382177	620	KP001174	16,758
	SRR765710	2 × 101	0.294%				

2.2 Automatic Selection of Bootstrap Sample Size

The number of read pairs in a bootstrap sample has a significant effect on the quality of resulting assembly. Too small a number of reads may produce fragmented assemblies due to lack of coverage for some regions. Too large a number may be detrimental by increasing the complexity of the assembly graph and making it more difficult to remove tangles generated by sequencing errors. In the original version of SMART [5] the number of read pairs in a bootstrap sample is specified by the user, and this can lead to many trial-and-error runs to find the optimal coverage.

In SMART2 we implemented a simple doubling strategy for automatically selecting the number of read pairs used in each bootstrap sample. Based on SMART experiments with manually specified numbers of read pairs we noted that a mean read coverage of the mitochondrial genome between 20× and 40× generates complete mitogenomes with high success rate. Unfortunately, it is difficult to analytically estimate the number of read pairs that yields a mitochondrial coverage in this range since the percentage of mitochondrial reads in real WGS datasets can vary by orders of magnitude [3] and the exact sizes of the nuclear and mitochondrial genomes are often not known *a priori*. For a single WGS library, SMART2 starts with 100,000 read pairs and then iteratively doubles the number of pairs until reaching an estimated mean mitochondrial read coverage of 20× or more. For two WGS libraries, SMART2 uses a similar doubling strategy starting with 100,000 read pairs and stopping when the *sum* of the mean mitochondrial read coverages estimated from the two libraries is 20× or more.

3   Results and Discussion

3.1   Comparison of Coverage-Based Filters and Assembly Accuracy on WGS Datasets from Species with Published Mitogenomes

For a detailed assessment, including evaluating the effectiveness of the SMART2 coverage-based filters and comparing assembly accuracy with previous methods we used five two-library datasets from species with published mitogenomes. The datsets are comprised of three insects (*Anopheles stephensi*, *Anopheles funestus*, and *Drosophila mauritiana*) and two birds (*Parus major* and *Pseudopodoces humilis*). Accession numbers and basic statistics for the five datasets are provided in Table 1.

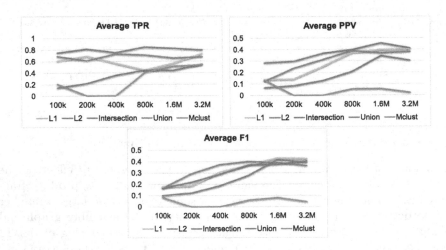

Fig. 4. Accuracy of single and multi-library coverage-based filters on 100k–3.2M read pairs randomly selected from libraries in Table 1.

Figure 4 plots the *True Positive Rate* (TPR), *Positive Predictive Value* (PPV), and F1 score (harmonic mean of TPR and PPV) achieved by the MCLUST, Union, and Intersection filters of SMART2 as the total number of read pairs is varied between 100k and 3.2M. All values are averages over the five species in Table 1. For comparison we include the average TPR, PPV, and F1 score of single library filters (L1 and L2). The results underscore the poor performance of the 2-dimensional mixture model (MCLUST), and the different tradeoffs achieved between TPR and PPV by the Union and Intersection filters. Specifically, for a fixed number of reads, the Union filter typically achieves a higher TPR but lower PPV than single library filters, while the Intersection filter does the opposite. In these experiments, the Intersection filter yields an F1 score comparable with single library filters for the lower range of tested number of read pairs, but both the Union and Intersection filters converge towards the performance of single library filters as the number of read pairs exceeds one million.

Assembly accuracy results generated by SMART2 and three other tools (Norgal [2], NOVOPlasty [11], and PlasmidSPAdes [7]) on the datasets described in Table 1 are given in Table 2. The number of read pairs used for assembly, indicated in last column for both single and two-library runs, was selected using the doubling strategy implemented described in Sect. 2. For each method, the assembled sequence length and percentage identity to the published reference are typeset in bold when the reconstructed sequence is a circular genome.

On all datasets Norgal failed to generate any contigs or generated nuclear rather than mitochondrial contigs, consistent with the poor performance reported for low-coverage WGS data in [3]. NOVOPlasty generated circular mitogenomes from two of the ten libraries, but failed on one library, and generated only incomplete mitogenomes from the remaining seven. PlasmidSPAdes

Table 2. Assembled sequence length and percentage identity to the published reference for low-coverage WGS datasets from species with published mitogenomes. Numbers in bold indicate a complete circular mitogenome.

Species	Library	Norgal	NOVOPlasty	PlasmidSPAdes	SMART2	#Pairs
A. stephensi	SRR630623	Nuclear	2,962	14,974	15,153	800k
			66.6%	99.5%	99.6%	
	SRR630669	Nuclear	2,428	15,324	**15,412**	800k
			99.8%	99.5%	**99.5**	
	Both	N/A	N/A	N/A	15,283	2 × 400k
					99.8%	
A. funestus	SRR630620	–	2,105	12,819	13,424	800k
			99.6%	41.6%	99.5%	
	SRR630619	–	2,402	15,176	13,369	800k
			99.4%	99.5%	99.4%	
	Both	N/A	N/A	N/A	10,502	2 × 400k
					99.5%	

(continued)

Table 2. (*continued*)

Species	Library	Norgal	NOVOPlasty	PlasmidSPAdes	SMART2	#Pairs
D. mauritiana	SRR1560275	Nuclear	14,922 99.9%	15,411 96.5%	15,462 96.7%	400k
	SRR1560276	Nuclear	9,327 99.9%	15,245 97.9%	15,643 95.3%	400k
	Both	N/A	N/A	N/A	15,397 97%	2 × 200k
P. major	SRR2961765	–	**16,774 99.8%**	**16,791 99.7%**	**16,814 99.6%**	1,6M
	SRR2961767	Nuclear	**16,774 99.8%**	**16,790 99.7%**	**16,813 99.6%**	1.6M
	Both	N/A	N/A	N/A	**16,814 99.6%**	2 × 800k
P. humilis	SRR765709	Nuclear	–	16,852 98.8%	**16,797 99.1%**	1.6M
	SRR765710	–	8,139 99.5%	**16,774 99.3%**	**16,797 99.1%**	800k
	Both	N/A	N/A	N/A	**16,797 99.1%**	2 × 400k

generated circular mitogenomes from three of the ten libraries, while SMART2 succeed on five of the ten single-library runs and two of the five two-library runs.

3.2 SMART2 Assembles Novel Mitogenomes

In a pilot project to assemble "orphan mitogenomes" for species with publicly available WGS data but no published mitogenome sequence we ran SMART2 on WGS datasets from 18 mammals (*Abrocoma cinerea, Arvicola amphibius, Babyrousa babyrussa, Canis rufus, Coendou bicolor, Cratogeomys planiceps, Ctenodactylus gundi, Cuniculus paca, Grammomys surdaster, Heteromys oasicus, Hippotragus niger kirkii, Hippotragus niger niger, Pipistrellus pipistrellus, Pusa hispida saimensis, Rhacophorus chenfui, Sciurus carolinensis, Sorex palustris*, and *Urocitellus parryii*) and 8 amphibians (*Agalychnis moreletii, Brachycephalus ferruginus, Brachycephalus pombali, Cycloramphus boraceiensis, Hyla arborea, Hylodes phyllodes, Melanophryniscus xanthostomus*, and *Oophaga pumilio*). Basic information about the 26 datasets is given in Table 3. The number of read pairs was selected automatically using the doubling strategy for all datasets except *A. cinerea, G. surdaster*, and *H. arborea*, for which we manually increased the number of read pairs after automatic selection failed to assemble complete circular genomes.

Table 3. WGS datasets from 26 metazoans without published mitogenomes. mtDNA content was estimated by aligning the reads against the SMART2 assembly only when the latter was a complete sequence. The number of read pairs was selected automatically by using the doubling strategy described in Sect. 2 except for the three species marked with a dagger for which it was manually increased after automatic selection failed to assemble a complete circular mitogenome. A "*" indicates datasets for which all available read pairs were used.

Species	Run ID	Read length	% mtDNA	#Pairs Used	Seed ID
Abrocoma cinerea	SRR8885043	2 × 151	1.490	2,000,000†	AF244388
Agalychnis moreletii	SRR8327212	2 × 182	NA	1,600,000	EF125031
Arvicola amphibius	ERR3316036	2 × 151	0.002	51,200,000	LT546162
Babyrousa babyrussa	ERR2984475	2 × 100	0.022	12,800,000	AY534302
Brachycephalus ferruginus	SRR5837605	2 × 251	NA	856,599*	HQ435708
Brachycephalus pombali	SRR5837604	2 × 251	NA	846,282*	HQ435714
Canis rufus	SRR8066613	2 × 101	0.565	400,000	U47043
Coendou bicolor	SRR8885018	2 × 151	3.372	100,000	U34852
Cratogeomys planiceps	SRS4613652	2 × 151	2.537	100,000	AY545541
Ctenodactylus gundi	SRR8885020	2 × 151	0.246	400,000	U67301
Cuniculus paca	SRS4613635	2 × 151	0.371	400,000	JF459150
Cycloramphus boraceiensis	SRR4019528	2 × 305	NA	1,776,547*	KU494395
Grammomys surdaster	SRS4524074	2 × 151	0.689	10,000,000†	KY753991
Heteromys oasicus	SRR8885041	2 × 151	0.965	200,000	ABCSA423-06
Hippotragus niger kirkii	SRS4184270	2 × 101	0.017	25,600,000	AF049388
Hippotragus niger niger	SRR8366604	2 × 101	0.012	51,200,000	AF049393
Hyla arborea	SRR2157967	2 × 101	NA	10,000,000†	JN312692
Hylodes phyllodes	SRR4019434	2 × 305	NA	1,055,455*	DQ502873
Melanophryniscus xanthostomus	SRR5837589	2 × 251	NA	977,403*	KX025607
Oophaga pumilio	SRR7627571	2 × 49	NA	3,200,000	KX574023
Pipistrellus pipistrellus	ERR3316150	2 × 151	0.007	25,600,000	HM380206
Pusa hispida saimensis	ERR2608991	2 × 170	0.098	1,600,000	JX109798
Rhacophorus chenfui	SRR5248583	2 × 300	NA	3,477,603*	KP996818
Sciurus carolinensis	ERR3312500	2 × 151	2.791	100,000	JF457099
Sorex palustris	SRR8451745	2 × 150	NA	6,400,000	MG421461
Urocitellus parryii	SRR8263911	2 × 151	0.609	200,000	KX646821

As shown in Table 4, out of the 26 datasets, SMART2 generated 15 complete circular mitogenomes and 11 partial mitogenomes, for a total of 403,541 bp. NOVOPlasty and PlasmidSPAdes generate only 5 and 1 complete circular mitogenomes, respectively. As seen in Fig. 5, when all three methods succeed, agreement between the assembled sequences is very high. However, NOVOPlasty and PlasmidSPAdes have a much higher failure rate than SMART2, generating a total of only 258,538 bp and 224,818 bp of mitogenomic sequences, respectively.

To further assess the accuracy of mitogenomes assembled by SMART2 we performed a joint phylogenetic analysis with published complete mitogenome sequences of up to two species in the same family, whenever the latter could be identified (see Table 4 for accession numbers). The joint phylogeny annotated using iTOL [17] is shown in Fig. 6. The phylogeny was constructed using Fast-Tree [20] with 10,000 bootstraps and the jModelTest [10] model of sequence

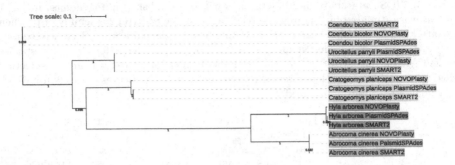

Fig. 5. Phylogenetic tree of mitogenomes assembled by SMART2, NOVOPlasty, and PlasmidSPAdes.

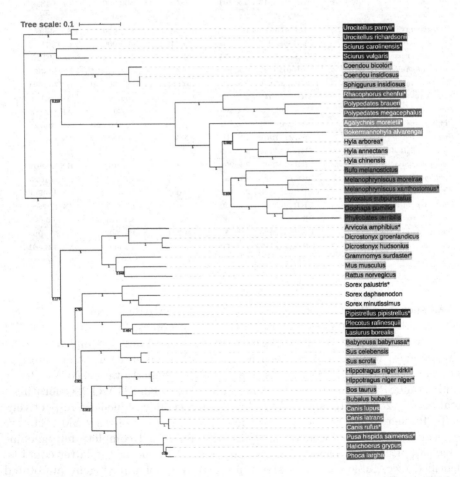

Fig. 6. Phylogenetic tree comparing SMART2 mitogenomes with published mitogenomes of related species.

Table 4. Mitochondrial sequences assembled for 26 metazoans without previously published mitogenomes. Sequence lengths typeset in bold indicate complete circular mitogenomes. Up to two complete mitogenomes from species in the same family were used for phylogenetic validation.

Species	Related Species	GenBank Accession	NOVO Plasty	Plasmid SPAdes	SMART2	TPA ID	Protein Coding	tRNA	rRNA
A. cinerea	NA	NA	17,047	16,863	**16,759**	BK010962	13	22	2
A. moreletii	Bokermannohyla alvarengai	NC_036493	12,236	18,107	15,781	BK010959	13	22	2
A. amphibius	Dicrostonyx groenlandicus	NC_034313	291	Failed	**16,359**	BK010955	13	22	2
	D. hudsonius	NC_034307							
B. babyrussa	Sus scrofa	MK251046	10,737	Nuclear	**16,645**	BK010954	13	22	2
	Sus celebensis	KJ789952							
B. ferruginus	NA	NA	6,846	16,135	9,806	BK010965	11	8	0
B. pombali	NA	NA	Failed	No match	9,847	BK010964	11	8	0
C. rufus	Canis lupus	KF857179	9,149	16,585	**16,474**	BK011186	13	22	2
	Canis latrans	NC_008093							
C. bicolor	Coendou insidiosus	NC_021387	16,630	16,771	**16,687**	BK010970	13	22	2
	Sphiggurus insidiosus	JX312693							
C. planiceps	NA	NA	16,438	16,682	**16,534**	BK010972	13	22	2
C. gundi	NA	NA	264	16,315	**16,101**	BK010971	13	22	2
C. paca	NA	NA	16,719	No match	**16,627**	BK010958	13	22	2
C. boraceiensis	NA	NA	16,933	No match	15,654	BK010967	13	22	2
G. surdaster	Mus musculus	KY018919	9,398	Nuclear	**16,308**	BK010969	13	22	2
H. oasicus	Rattus norvegicus	KM577634	16,401	7,655	**16,401**	BK010960	13	22	2
	NA								
H. niger kirkii	Bos taurus	AF492351	16,508	Timeout	**16,508**	BK011057	13	22	2
	Bubalus bubalis	MK234704							
H. niger niger	Bos taurus	AF492351	Timeout	Failed	**16,506**	BK011056	13	22	2
	Bubalus bubalis	MK234704							

Continued on next page

Table 4. (*continued*)

Species	Related Species	GenBank Accession	NOVO Plasty	Plasmid SPAdes	SMART2	TPA ID	Protein Coding	tRNA	rRNA
H. arborea	Hyla annectans	NC_025309	16,757	15,958	15,751	BK010919	13	22	2
	Hyla chinensis	NC_006403							
H. phyllodes	NA	NA	16,037	17,261	10,479	BK010968	11	13	0
M. xanthostomus	Melanophryniscus moreirae	NC_037378	5,000	16,670	15,953	BK010963	13	22	2
	Bufo melanostictus	NC_005794							
O. pumilio	Phyllobates terribilis	NC_037380	Failed	No match	15,856	BK010961	13	22	2
	Hyloxalus subpunctatus	NC_037379							
P. pipistrellus	Lasiurus borealis	NC_016873	252	Failed	16,458	BK010957	13	22	2
	Plecotus rafinesquii	NC_016872							
P. hispida saimensis	Phoca largha	FJ895151	258	16,667	16,499	BK011058	13	22	2
	Halichoerus grypus	NC_001602							
R. chenfui	Polypedates braueri	NC_042797	14,999	Nuclear	14,441	BK010966	12	22+1dup	2
	Polypedates megacephalus	AY458598							
S. carolinensis	Urocitellus richardsonii	NC_031209	7,025	16,610	16,537	BK010956	13	22	2
	S. vulgaris	NC_002369							
S. palustris	Sorex daphaenodon	NC_044107	16,151	Nuclear	16,108	BK011027	13	22	2
	Sorex minutissimus	NC_042196							
U. parryii	Urocitellus richardsonii	NC_031209	16,462	16,539	16,462	BK011059	13	22	2
	Sciurus vulgaris	NC_002369							

evolution from a multiple alignment generated using MAFFT [15]. The phylogeny places the sequences of each family within independent clades, supporting the accuracy of SMART2 assemblies. Assembly accuracy is further supported by the completeness of MITOS annotations (see Table 4 for the number of annotated genes for each species). All mtDNA sequences assembled by SMART2 for the 26 species in the pilot project have been submitted to GenBank as Third Party Annotation (TPA) sequences (see Table 4 for TPA accession numbers).

4   Conclusions

In this paper we presented SMART2, an enhanced pipeline that can assemble high quality mitochondrial genomes from low coverage WGS datasets with minimal user intervention. SMART2 succeeded in generating mitochondrial sequences – including 15 complete circular mitogenomes – for 26 metazoan species with WGS data but no previously published mitogenomes in NCBI databases. An additional complete mitogenome assembled using the multilibrary feature of SMART2 will be published separately [4]. The SMART2 pipeline is publicly available via a user-friendly Galaxy interface at https://neo.engr.uconn.edu/?tool_id=SMART2.

References

1. Afgan, E., et al.: The Galaxy platform for accessible, reproducible and collaborative biomedical analyses: 2018 update. Nucleic Acids Res. **46**(W1), W537–W544 (2018)
2. Al-Nakeeb, K., Petersen, T.N., Sicheritz-Pontén, T.: Norgal: extraction and de novo assembly of mitochondrial DNA from whole-genome sequencing data. BMC Bioinform. **18**(1), 510 (2017)
3. Alqahtani, F., Mandoiu, I.: Statistical mitogenome assembly with repeats. J. Comput. Biol. online ahead of print (2020). https://doi.org/10.1089/cmb.2019.0505
4. Alqahtani, F., Duckett, D., Pirro, S., Măndoiu, I.I.: Complete mitochondrial genome of water vole, Microtus richardsoni (2020). In preparation
5. Alqahtani, F., Măndoiu, I.I.: Statistical mitogenome assembly with repeats. In: 8th IEEE International Conference on Computational Advances in Bio and Medical Sciences (2018)
6. Alves-Silva, J., et al.: The ancestry of Brazilian mtDNA lineages. Am. J. Hum. Genet. **67**(2), 444–461 (2000)
7. Antipov, D., Hartwick, N., Shen, M., Raiko, M., Lapidus, A., Pevzner, P.A.: plasmidSPAdes: assembling plasmids from whole genome sequencing data. Bioinformatics **32**(22), 3380–3387 (2016)
8. Calabrese, C., et al.: MToolBox: a highly automated pipeline for heteroplasmy annotation and prioritization analysis of human mitochondrial variants in high-throughput sequencing. Bioinformatics **30**(21), 3115–3117 (2014)
9. Cochrane, G., et al.: Evidence standards in experimental and inferential INSDC third party annotation data. OMICS J. Integr. Biol. **10**(2), 105–113 (2006)
10. Darriba, D., Taboada, G.L., Doallo, R., Posada, D.: jModelTest 2: more models, new heuristics and parallel computing. Nat. Methods **9**(8), 772 (2012)

11. Dierckxsens, N., Mardulyn, P., Smits, G.: NOVOPlasty: de novo assembly of organelle genomes from whole genome data. Nucleic Acids Res. **45**(4), e18–e18 (2016)

12. Gupta, A., Bhardwaj, A., Sharma, P., Pal, Y., et al.: Mitochondrial DNA-a tool for phylogenetic and biodiversity search in equines. J. Biodivers. Endangered Species **2015** (2015)

13. Hahn, C., Bachmann, L., Chevreux, B.: Reconstructing mitochondrial genomes directly from genomic next-generation sequencing reads–a baiting and iterative mapping approach. Nucleic Acids Res. **41**(13), e129–e129 (2013)

14. Hebert, P.D., Ratnasingham, S., de Waard, J.R.: Barcoding animal life: cytochrome c oxidase subunit 1 divergences among closely related species. Proc. Roy. Soc. London Ser. B Biolog. Sci. **270**(suppl_1), S96–S99 (2003)

15. Katoh, K., Misawa, K., Kuma, K.I., Miyata, T.: MAFFT: a novel method for rapid multiple sequence alignment based on fast Fourier transform. Nucleic Acids Res. **30**(14), 3059–3066 (2002)

16. Kurabayashi, A., Sumida, M.: Afrobatrachian mitochondrial genomes: genome reorganization, gene rearrangement mechanisms, and evolutionary trends of duplicated and rearranged genes. BMC Genom. **14**(1), 633 (2013)

17. Letunic, I., Bork, P.: Interactive tree of life (iTOL) v4: recent updates and new developments. Nucleic Acids Res. **47**(W1), W256–W259 (2019)

18. Li, W.X., et al.: The complete mitochondrial dna of three monozoic tapeworms in the caryophyllidea: a mitogenomic perspective on the phylogeny of eucestodes. Parasites Vectors **10**(1), 314 (2017)

19. Melton, T., Holland, C., Holland, M.: Forensic mitochondria DNA analysis: current practice and future potential. Forensic Sci. Rev. **24**(2), 101 (2012)

20. Price, M.N., Dehal, P.S., Arkin, A.P.: Fasttree: computing large minimum evolution trees with profiles instead of a distance matrix. Mol. Biol. Evol. **26**(7), 1641–1650 (2009)

21. Ratnasingham, S., Hebert, P.D.: BOLD: The barcode of life data system (http://www.barcodinglife.org). Mol. Ecol. Notes **7**(3), 355–364 (2007)

22. Scrucca, L., Fop, M., Murphy, T.B., Raftery, A.E.: mclust 5: clustering, classification and density estimation using Gaussian finite mixture models. R J. **8**(1), 205–233 (2016)

23. Trevisan, B., Alcantara, D.M., Machado, D.J., Marques, F.P., Lahr, D.J.: Genome skimming is a low-cost and robust strategy to assemble complete mitochondrial genomes from ethanol preserved specimens in biodiversity studies. PeerJ **7**, e7543 (2019)

24. Veltri, K.L., Espiritu, M., Singh, G.: Distinct genomic copy number in mitochondria of different mammalian organs. J. Cell. Physiol. **143**(1), 160–164 (1990)

Author Index

Printed in the United States
By Bookmasters